The Eugenic Fortress

CEU Press Studies in the History of Medicine

Volume VII

Series Editor: Marius Turda

Published in the series:

Svetla Baloutzova
Demography and Nation
Social Legislation and Population Policy in Bulgaria, 1918–1944

Christian Promitzer · Sevasti Trubeta · Marius Turda, eds.
Health, Hygiene and Eugenics in Southeastern Europe to 1945

Francesco Cassata
Building the New Man
Eugenics, Racial Science and Genetics in Twentieth-Century Italy

Rachel E. Boaz
In Search of "Aryan Blood"
Serology in Interwar and National Socialist Germany

Richard Cleminson
Catholicism, Race and Empire
Eugenics in Portugal, 1900–1950

Maria Zarimis
Darwin's Footprint
Cultural Perspectives on Evolution in Greece (1880–1930s)

The Eugenic Fortress

The Transylvanian Saxon Experiment in Interwar Romania

Tudor Georgescu

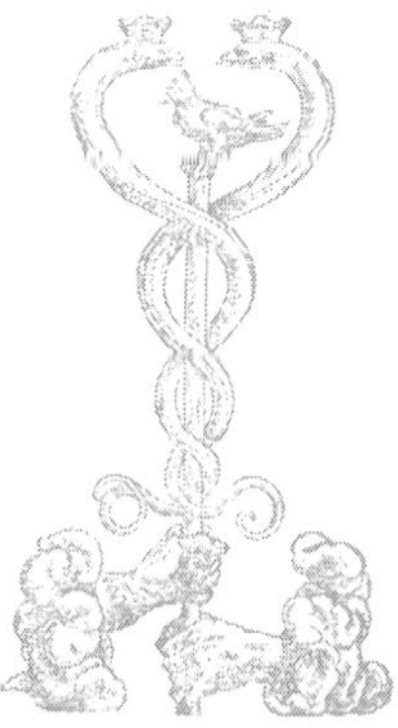

CEU PRESS

Central European University Press
Budapest—New York

Published in 2016 by
Central European University Press
An imprint of the
Central European University Limited Liability Company
Nádor utca 11, H-1051 Budapest, Hungary
Tel: +36-1-327-3138 or 327-3000 · *Fax*: +36-1-327-3183
E-mail: ceupress@press.ceu.edu
Website: www.ceupress.com

224 West 57th Street, New York NY 10019, USA
E-mail: meszarosa@press.ceu.edu

ISBN 978-963-386-139-4

ISSN 2079-1119

Library of Congress Cataloging-in-Publication Data

Names: Georgescu, Tudor, author.
Title: The eugenic fortress : the Transylvanian Saxon experiment in interwar Romania / Tudor Georgescu.
Description: New York : Central European University Press, 2016. | Series: CEU Press studies in the history of medicine, ISSN 2079-1119 ; Volume VII | Includes bibliographical references and index. | Description based on print version record and CIP data provided by publisher; resource not viewed.
Identifiers: LCCN 2015049638 (print) | LCCN 2015043867 (ebook) | ISBN 9789633861417 (pdf) | ISBN 9789633861394 (hardcover)
Subjects: LCSH: Eugenics—Romania—History—20th century. | Fascism—Romania—History—20th century. | Germans—Romania—Transylvania—History—20th century.
Classification: LCC HQ755.5.R6 (print) | LCC HQ755.5.R6 G46 2016 (ebook) | DDC 363.9/209498—dc23

LC record available at http://lccn.loc.gov/2015049638

Printed by Prime Rate Kft., Hungary

CONTENTS

LIST OF ILLUSTRATIONS

ACKNOWLEDGEMENTS

First of all, I owe the Wellcome Trust a world of gratitude for its kind and generous support of my doctoral research that eventually led to this book. But I am eternally indebted to my beloved wife Sabine, whose obviously boundless patience with an increasingly erratic husband has never failed to amaze me. Nor, needless to say, my awesome whirlwinds Dennis and Anja—I'm sure you'll read and love this when you're old enough… I am particularly thankful to be blessed with parents whose limitless love and trust have been such an inspiration and source of strength. I would also like to thank Marius Turda for an incredible journey, cherished memories, and a heartfelt friendship.

I would also like to thank, again and again, Monica Vlaicu, with whom I've had the pleasure of spending many an afternoon rooting through archival treasures at the Teutsch House. Similarly, I am very grateful indeed to Alexiu Tatu at the Romanian National Archives, who kindly helped me use the incredible files his archive holds and who has been a constant source of support. Thank you, too, to the many friends and colleagues who helped in all manners, for indulging the many long-winded ramblings about eugenics, for being the best imaginable companions on academia's winding road.

ABBREVIATIONS

BSV – Land Preservation Society (Bodenschutzverein)

DVR – National Party of the Germans in Romania (Deutsche Volkspartei in Rumänien)

FDV – Research Institute of the German National Community in Romania (Forschungsinstitut der Deutschen Volksgruppe in Rumänien)

HJS – *Half-Yearly Journal for Southeast European History, Literature, and Politics* (*Halbjahresschrift für Südosteuropäische Geschichte, Literatur und Politik*)

IOGT – International Order of Good Templars

ISB – Institute for Statistics and Population Policy of the German National Community in Romania (Institut für Statistik und Bevölkerungspolitik der Deutschen Volksgruppe in Rumänien)

LK – Governing Council of the Protestant National Church (Landeskonsistorium der evangelischen Landeskirche Augsburger Bekentnisses in Rumänien)

LSBS – National Department for Statistics, Population Policy, and Genealogy (Landesarbeitsstelle für Statistik, Bevölkerungspolitik und Sippenwesen)

LSS – National Office for Statistics and Genealogy (Landesamt für Statistik und Sippenwesen)

NA Sibiu – Romanian National Archives, Sibiu Branch (Arhivele Naţionale ale României, Direcţia Sibiu)

NEDR – National Movement of Renewal of the Germans in Romania (Nationale Erneuerungsbewegung der Deutschen in Rumänien)

NSDAP der VgrDR – NSDAP of the German National Community in Romania (NSDAP der Deutschen Volksgruppe in Rumänien)

NSDR – National Socialist Self-Help Movement of the Germans in Romania (Nationalsozialistische Selbsthilfebewegung der Deutschen in Rumänien)

OB – *East German Observer* (*Ostdeutscher Beobachter*)

SDT – *Transylvanian German Daily* (*Siebenbürgisch-Deutsches Tageblatt*)

VdK – Society of Child Enthusiasts (Verein der Kinderfrohen)

VDR – Association of Germans in Romania (Verband/Volksgemeinschaft der Deutschen in Rumänien)

VgrDR – German National Community in Romania (Volksgruppe der Deutschen in Rumänien)

ZAEV – Central Archive of the National Protestant Church A.B. in Romania, Sibiu (Zentralarchiv des Landeskonsortiums der Evangelischen Kirche A.B. in Rumänien, Hermannstadt)

ZfSL – *Journal for Transylvanian Regional Studies* (*Zeitschrift für Siebenbürgische Landeskunde*)

INTRODUCTION

The post-Versailles reshuffle of Eastern Europe's borders signified the onset of a new era in the relationship between minorities and host nations, old and new alike. It marked the beginning of an interwar period awash with homogenizing nation-building projects that sought meaning and purpose in the new geopolitical realities that fostered them. Romania, for one, could certainly be content with the rewards reaped at the negotiating tables of a waning First World War belatedly joined on the Entente's side on August 27, 1916. All the more so considering that following a short burst of territorial conquests in southeastern Hungary, Romania had been overrun by Austro-German forces, which captured Bucharest four months later. However, with the harsh Treaty of Bucharest annulled by Germany's defeat, Romania's fortunes were dramatically reversed. The Treaty of Saint-Germain with Austria awarded Romania Bukovina, while that of Trianon with Hungary swelled Romania's territorial gains with Transylvania.[1] A third peace accord, with Bulgaria, granted Romania Dobrudja, while the Bessarabian National Congress had already voted to separate from Bolshevik Russia. Although the acquisition of Bessarabia was of significant economic and strategic value, Transylvania was the main prize, the ultimate fulfilment of nationalist dreams and their "total nation-state" ambitions. Irina Livezeanu likened it to a national "revolution," while emphasizing that the integra-

1 The geographic concept of "Transylvania" is itself fairly fluid, and must briefly be outlined. The Trianon Treaty had not merely awarded Romania the historic Transylvania (the medieval *Voievodat*), but large portions of eastern Hungary that harbored significant Romanian populations as well—namely, Crişana, Satu-Mare, Maramureş, and part of the Banat. It is this territorial conglomerate that is commonly understood to mean Transylvania. In this study, the term is used in its historic remit, and refers to Siebenbürgen alone.

tion of regions with such divergent cultural and linguistic traditions made for a thoroughly turbulent nation-building process.[2] Greater Romania thus entered the interwar era drastically swollen territorially, and with a remarkably heterogeneous assortment of newly acquired ethnic minorities—the second largest group of which encompassed the various German enclaves.

The Transylvanian Saxons were one of the oldest and largest of the German communities to have found themselves holding Romanian passports in 1920, having settled in the region in the thirteenth century.[3] In contrast to most other German minorities in Romania, the roughly 250,000 Saxons had a firmly entrenched sense of national identity, strong urban and rural economies, established parliamentary and local political traditions, and a virtually omnipresent national Protestant church hierarchy (Evangelische Landeskirche Augsburger Bekentnisses), entrusted with the Saxon school system.[4] Despite adamant assurances by its elites that these pillars of Saxon identity were historical fixtures, the realities of the geopolitical storm gathering over Europe; the flurry of new ideas on, and definitions of nationhood and race rapidly replacing cultural and linguistic paradigms; and the growing pressures exerted by an increasingly nationalist Romanian nation-building project wreaked havoc on the Saxon economic, social, and political life. To study the shifts that the Saxon sense of self underwent in interwar Romania—namely, how it adopted and adapted a eugenically framed redefinition of Saxon national identity—offers a particularly enticing case study of an ethnic minority's embrace of eugenics in the name of national salvation.

It is peculiar that much of the historiography on the Transylvanian Saxons simply bypasses the eugenic discourse as it emerged and evolved in the early twentieth century.[5] The omission is all the more surprising considering the significant influence that eugenic population policies eventually came to exert.

---

2 Livezeanu, *Cultural Politics in Greater Romania*, 7.

3 For a wider overview of Saxon history, see Roth, *Kleine Geschichte Siebenbürgens*; Wagner, *Geschichte der Siebenbürger Sachsen*; Gündisch and Beer, *Siebenbürgen und die Siebenbürger Sachsen*; and Schuster, ed., *Epoche der Entscheidungen*.

4 For an overview of the church's interwar history, see Binder, "Die Evangelische Landeskirche A.B."; König, ed., *Siebenbürgen zwischen den beiden Weltkriegen*, 237–264; Böhm and Braeg, eds., *Dr. Viktor Glondys*; Möckel, *Umkämpfte Volkskirche*; Wien, *Kirchenleitung über dem Abgrund*; Wien, "'Ich rufe alle auf'"; and Wien, "'Entjudung' und Nationalsozialismus."

5 For instance, Szelényi, "From Minority to *Übermensch*." An otherwise excellent piece, Szelényi's article could have taken its arguments about "proto-Nazi" or "organic National Socialism" one step further.

In broad strokes, the Transylvanian Saxon experiment with eugenics underwent three distinct phases between 1885 and 1944: a formative period between 1885 and 1918; a period of increasing politicization, radicalization, and piecemeal implementation unto 1940; and a third period, in the Third Reich's service, between 1940 and 1944. While this investigation primarily focuses on the second period, specifically on the theoretical and institutional grounding of eugenic thought in the 1920s and 1930s, it is nonetheless essential to outline, if only briefly, the mindset from which it emerged, and the reality by which it was eventually superseded.

The first, formative phase, spanning the years 1885 to 1918, was essentially dominated by two competing evaluations of Saxon demography. When Oskar von Meltzl published his *The Statistics of the Saxon Rural Population*[6] to much acclaim in 1885, the underlying "destruction theory" (*Vernichtungslehre*) employed to account for Saxon territorial and population losses was unequivocally canonized by the Saxon status quo. In essence, Meltzl explained the declining number of Saxon settlements as the result of Hungarian and Romanian populations moving into areas emptied of their Saxon inhabitants by wars and plagues, rather than attributing their depletion to internal (demographic) processes or the competition for *Lebensraum* between Transylvania's various ethnicities. In fact, Meltzl believed the exact opposite to be the case, when in a truly Malthusian manner he asserted that while these ethnic "others" might enjoy higher birthrates statistically, Saxon dominance of land ownership would place very finite limits on their potential for further expansion, and hence on the likelihood of them "becoming a threat." Reflecting a self-perception of Saxon superiority, Meltzl's "optimistic" assessment of continued Saxon domination came to define Saxon public opinion and policy making for a good three decades thereafter. What is more, the status of the "destruction theory" as historic truth gave an emerging eugenic discourse its initial impetus and purpose: to substantiate its sense of an impending crisis born of internal, rather than external processes, such as declining birthrates and a creeping racial degeneration that not only weakened the Saxon nation's ability to resist, but, which was worse, allowed (and even furthered) their Romanian rivals' "conquest" of Saxon *Lebensraum*.

6 Meltzl, "Statistik der Sächsischen Landbevölkerung in Siebenbürgen."

The alternative model, insisting on "displacement" rather than replacement, was inexhaustibly propagated by Heinrich Siegmund, the singularly most important and prolific Saxon eugenicist. The first to truly cast the Saxon nation as a biological entity embroiled in a battle for survival, Siegmund published a remarkable article, "On Transylvanian Saxon Racial Hygiene," as early as 1901.[7] In it he called for a concerted engagement with a burgeoning field that "can serve to help strengthen our Saxon nation as a biological unit, as a race, to adapt to the environment and so safeguard its future." He was quick to point to the potential advantages racial hygiene offered, and complained that it was hardly ever discussed, although "the conscious control of variation, despite being an incredibly important weapon, is indeed perhaps one of the most important ones in the battle to preserve our national peculiarity [*Eigenart*]."[8] Through this biologization of the Saxon nation, or rather its medicalization, Siegmund fashioned a "science of national defense" through which eugenicists would diagnose the relative "health" of any particular Saxon settlement. The first phase of the wider Saxon eugenic discourse is hence deeply indebted to Siegmund's conceptual and statistical work on the core questions of why the Saxons seemed to be losing out both numerically and territorially, and how racial hygiene could salvage the Saxon nation biologically in its battle for survival.

The second phase of the eugenic discourse began with Transylvania's union with Romania, and spanned the interwar years, 1919 to 1940. Most importantly, it witnessed the emergence of a new, increasingly nationalist generation of Saxon eugenicists. The emphasis was placed on the theoretical honing, institutional grounding, and most importantly, the political pursuit of population policies geared towards the rejuvenation of an idealized Saxon human capital. This process, in which "quality" consistently overrode "quantity," was substantially furthered by the emergence of an indigenous Saxon fascist movement for renewal, Fritz Fabritius's Self-Help. While the nature of Saxon fascism in general, and its involvement with German National Socialism in particular, are highly contested fields of study, the Self-Help's embrace of eugenic thought is as unequivocal and astonishing as its metamorphosis from book club to empowered regime within a decade of its appearance.

7 See Siegmund, *Zur sächsischen Rassenhygiene.*

8 Ibid., 9.

The Self-Help was initially created in Hermannstadt in 1922 by ten founding members, with the purpose of acquiring land and loans, particularly for the urban poor. By this time, Saxon public opinion was increasingly turning into opposition to an unreceptive political status quo, and the Self-Help always had a political brief above and beyond the economic policies. It was seen as a key solution to the Saxon dysgenic crisis, such as when we find Fabritius explaining that a "body that is sick all over cannot be healed in just one place. And so the purely economic self-help work is but a side project. Because the economy must necessarily heal if the nation does so from the inside out. The self-help work is much more than collecting and handing out money—it is a battlefront [*Kampfgemeinschaft*]."[9]

The Self-Help represented an indigenous fascist movement that adopted the National Socialist franchise in 1932, and was subsequently rebranded the National Socialist Self-Help Movement of the Germans in Romania (Nationalsozialistische Selbsthilfebewegung der Deutschen in Rumänien, or NSDR). Undoubtedly boosted by—but surely not conditional upon—the Third Reich's ascension to power, the NSDR succeeded in pushing a more or less watered-down version of its national program (*Volksprogramm*) through the Saxon Constitutional Assembly (Sachsentag) in October 1933, and found itself in control of the Saxon National Council (Volksrat) a month later. Following the Romanian government's ban of the NSDR in November 1933, the party reemerged ten days later as the National Movement for Renewal of the Germans in Romania (Nationale Erneuerungsbewegung der Deutschen in Rumänien, or NEDR), before being banned again in July 1934. The absence of a party structure appears not to have compromised the movement's overall cohesion; on the contrary, Fabritius was elected chair of the Association of Germans in Romania (Verband der Deutschen in Rumänien, or VDR) in June 1935. The Self-Help had arrived at its final destination, political control over all German settlement areas in Romania, but with Fabritius's election immediately followed by an attempted coup, the next three years descended into bitter feuding between the movement's moderate and radical wings, a crisis eventually resolved by Berlin in 1938.

9 Report on the meeting of Self-Help's Hermannstadt Regional Branch on October 25, 1929, ZAEV, LK 103.90 (1929), p. 19.

More than ideological radicalization, the political turmoil of the 1920s and 1930s also had a significant impact on the viability of implementing eugenic policies above and beyond private initiatives. The Saxons effectively had two overarching, large-scale organizations: the Protestant Landeskirche and the political establishment. With the latter paralyzed by deadlock by the mid-1920s and Saxon fascism's gradual rise to power, the church remained the key avenue through which to promote and pursue eugenic policies, as seen with Siegmund and Csallner. Within this context, it is important to note that the church not only fulfilled a vital role spiritually but, perhaps more importantly still to Saxon eugenicists, was the minority's most significant employer and land owner. Its control of the Saxon educational system in particular meant that the church soon came under pressure to endorse eugenic ideas and policies. Yet the church's unique role in Saxon society was eroded during the 1930s by the revival of Saxon politics under the fascist aegis, with its calls for a more centralized, political leadership.

As with any political ideology, it is particularly difficult to quantify and qualify what—if any—real impact the Self-Help eventually exerted over individual lives above and beyond membership cards and uniformed marches. The centerpiece of the fascist bid to reinvent not only the Saxon nation but all of Romania's presumed Germans was Wilhelm Schunn's "national neighborhoods" (*völkische Nachbarschaften*, or NBs). The NBs are a fascinating example of a historic tradition reinvented and revolutionized to suit the fascist discourse, and also a truly remarkable case study of how an ethnic minority strove to bypass its host state in a quest to cast an ever-wider eugenic net over its body politic. If the various incarnations of Csallner's organizations marked his increasingly radical attempts to institutionalize eugenic values, Wilhelm Schunn's creation was tasked with carrying fascist eugenics into each and every Saxon home.

The national neighborhoods, as recreated in 1933, were community support networks operated outside of the church's direct sphere of influence (which, for its part, dominated their rural variants), and furnished fascist Saxon politics with a vital tool by which to enforce policy decisions at a local level. Their provision of financial aid on the one hand, and ability to exclude undesirables from its ranks on the other, made them a formidable force. Schunn was also a close friend of Csallner's, and the neighborhoods' policy of awarding the fiscally substantial "honorary gift" (*Ehrengeschenk*) in celebration of the birth of a racially valuable family's fourth or more child was deeply indebted to many of Csallner's core ideals. These gifts came with strings attached, and the money had to be invested in items "of enduring value," so as to both retain Saxon property in danger of being sold to ethnic "others," and even encourage the re-homogenization of a fractured Saxon *Lebensraum*. As these "gifts" were designed to benefit qualitatively valuable offspring, rather than encourage a purely quantitative population increase, the potential recipients were subjected to evaluations of their "hereditary quality" (while all renditions of the neighborhood guidelines insist on this, it is important to point out that it has been impossible to find anything but circumstantial evidence thereof). Schunn took great pride in reviving what he considered to be the Indo-Germanic ideal type of community life,[10] a system that turned a numerical weakness, in comparison to their ethnic neighbors, into an organizational strength. If nothing else, the rise of the neighborhoods and their ability to enforce eugenic policies signifies the greater trend at work here—namely, that the interwar eugenic discourse was to lean increasingly on political rather than clerical power to legislate and execute its agenda from the mid-1930s onwards.

With the onset of the third period of the Saxon eugenic discourse in 1940, the rules of engagement changed decisively. 1940 was a cataclysmic year that witnessed a series of geopolitical upheavals, including, amongst many more, Transylvania's geographic dismemberment, with the northern territories returned to an ever-irredentist Hungary following the Second Viennese Accord; the resettlement of German minorities out of Bessarabia, Bukovina, and Dobrudja; the Romanian state's recognition of its German minorities as a legal body; and the Third Reich assuming direct control over Saxon internal affairs by appointing Andreas Schmidt to the post of National Leader (*Volksgruppenführer*). Schmidt, for his part, immediately set about enforcing the NSDAP of the German National Community in Romania (NSDAP der Deutschen Volksgruppe in Rumänien, or NSDAP der VgrDR), and radically reconfiguring Saxon society, for example by stripping the church of most of its institutional influence—most significantly, of its schools and youth organizations. This third phase of Saxon eugenics lasted until the catastrophic collapse of August 1944, and marked a fundamental method-

10 Schunn, *Nachbarschaften*, 5–6.

ological and dogmatic departure from those indigenous qualities this study sets out to explore. Symptomatically, none of the eugenic movement's most influential figures survived this sea change, with Siegmund having died in 1937, Fabritius relieved from his post in 1939, Schunn retiring in 1940, and Csallner effectively stripped of power and sidelined by 1941.

This exploration of Saxon eugenics focuses on the life and work of three core figures, namely Heinrich Siegmund, Alfred Csallner, and Wilhelm Schunn, especially on their contributions to the eugenic discourse in the 1920s, and on its gradual institutionalization through the church and at the hands of the fascist Self-Help movement in the 1930s. However, this is just the tip of the iceberg, and much more could be said about almost all of the various rich lines of inquiry that the Saxon experiment with national renewal has to offer, such as the import and adaptation of race-hygienic thought and literature from Germany and Austria (be that via students returning to Transylvania or otherwise); the nature and extent of German financial commitments; the level of engagement with other German minorities (most notably in the Baltics); the reassessment of gender roles in light of eugenic idioms; the medical establishment's involvement in these and subsequent debates on the physician's role as the guardian of a biologically redefined Saxon identity; social hygienic measures combating a wide assortment of health concerns, ranging from alcoholism to tuberculosis; the ever-evolving and adapting role of the church, and its murky relationship with the eugenic discourses that courted it; and crucially, the very feasibility of an ethnic minority designing, implementing, and enforcing eugenic population policies in the absence of a fundamental precondition: an empowered nation-state.

I. Imagining a "Eugenic Fortress": Fascist Who and Eugenic What?

This study engages with two broader questions: First, did the Transylvanian Saxons produce a eugenic discourse in the interwar period? And second, how viable could such a discourse be, given their status as an ethnic minority? It sets out to identify and define a Saxon eugenic discourse in terms of its ideological imperatives and available methodological means, and the extent to which these changed with the Self-Help's rise to prominence. It is precisely this search for practical means to eugenic ends that makes the Transylvanian Saxon case study so unusual and tantalizing, above and beyond its ideological content and national context.

The history of European eugenic movements overwhelmingly focuses on eugenic projects proposed and empowered by nation-states; in doing so, it frequently overlooks the possibility of ethnic minorities pursuing independent or even competing nation (re)building agendas. This study aims both to offer a new case study to the blossoming scholarship on interwar eugenic movements, as well as to introduce one distinct from many of those already researched, by virtue of being fostered and furthered by an ethnic minority. The Saxon case study will hopefully help illuminate this fascinating window onto interwar eugenics, and offer a useful framework through which to investigate how ethnic minorities responded to, and in some cases even advanced the rise of biological determinism more widely.

Essentially, the analysis offered here seeks to understand the origins and trajectory of the interwar Saxon eugenic and fascist movements by locating them within the more reflective and comparative concepts of a eugenic and a generic fascist discourse. That is, it uses generic ideal types as a way of framing these movements conceptually, as well as to facilitate their relation to the much wider and more nebulous arena of their contemporary counterparts. Neither of these conceptual frameworks is fundamentally revised from the core definitions offered, in particular, by Marius Turda on eugenics[11] and Roger Griffin on fascism.[12] This investigation does not endeavor to design or offer a new "mininum"[13] or matrix for either (while hopefully profitably pointing out the obvious areas of overlap between them), all the more so as bookshelves abound with ever-more-theoretical models and competing one-sentence definitions that testify to the extensive and multifaceted historiography of fascism.[14] While all of these models have something to offer to specific case studies, only a few can be easily reconciled with each

11 See his two seminal monographs, *Modernism and Eugenics and Eugenics and Nation in Early 20th Century Hungary*, as well as Turda and Gillette, *Latin Eugenics in Comparative Perspective*.

12 See, in particular, his two key works, *Modernism and Fascism* and *The Nature of Fascism*.

13 See, for example, Nolte, *The Three Faces of Fascism*.

14 For an overview, see Iordachi, "Comparative Fascist Studies: An introduction." To choose but two titles from the canon, see Weber, *Varieties of Fascism*; and Larsen, Hagtvet, and Myklebust, *Who Were the Fascists*. For a wider collection of theories, see, for example, from amongst the variety of anthologies: Griffin and Feldman, eds., *Fascism*; Kallis, *The Fascism Reader*; Iordachi, ed., *Comparative Fascist Studies*; and Costa Pinto, ed., *Rethinking the Nature of Fascism*.

other, given the ideological elasticity of both eugenics and fascism, their ability to derive policies from and appeal to most corners of the political spectrum, and their clear propensity to transform and adapt over time to changing sociocultural, economic, and political realities. Taking a different path, this study aspires to increase accessibility to a hitherto neglected (or otherwise insufficiently researched) case study by way of locating it within the academic literature on fascism and eugenics and nuancing an analytical framework that has moved beyond its previous, binary focus on National Socialist Germany and Fascist Italy to become more inclusive of the plethora of distinct socioeconomic, cultural, and political traditions that gave rise to an equally diverse set of fascist movements.

That said, both eugenics and fascism steadfastly remain conceptually fuzzy, or at least heavily contested concepts, despite a process of amalgamation (and hopefully of consensus building) since the late 1990s. Perhaps also a victim of their remarkable popular currency, fascism in particular (and eugenics increasingly so) has become a holdall for a veritable menagerie of movements and ideologies, and it is all too easy to get somewhat lost in the jungle of pseudo-, para-, proto-, or apparently proper fascisms in terms of their ideologies, styles, and various degrees of actual empowerment.

Neither fascism nor eugenics is a monolithic, homogenous construct, but a cacophony of protestations about a perceived descent into degeneration and the corresponding biopolitical aspirations to remedy it by retooling the relationship between the individual, an organically defined national body, and an increasingly assertive state—or otherwise executive body—that negotiates them. Obviously oversimplified given the enormous wealth of literature, eugenics and fascism are hence in many ways understood here as conceptual kin amongst a family of ideologies that actively sought to redefine and regulate the nation's body politic for the purpose of an anthropological revolution, the beginning of a new age and a new man. Both were driven by the same ideological engine, and offered distinctly revolutionary ideologies that strove to verbalize and alleviate intense societal anxieties about the Saxons' perceived decent into the abyss of national extinction. Both sought to revolutionize the definition of what being Saxon actually meant by supplanting sociocultural paradigms with biological ones. Both sought to re-enchant public time and space, and thereby strove towards an alternative, purified modernity. Both strove towards the biological rejuvenation and regeneration of an idealized and re-homogenized national body. And, perhaps most importantly, they joined forces in the late 1920s in a combined bid to design and build an ethnically exclusive "eugenic fortress" behind whose race-hygienic walls a nation reinvented could re-entrench and protect its *Lebensraum* and perceived racial distinctiveness from ethnic "others."

Eugenics, in rather broad strokes, is understood as an ideology grounded in the biological evaluation of a nation redefined as a distinct and unique "organism" subject to the laws of natural selection and heredity. It embodies the dogmatic reaction to seismic shifts in a community's socioeconomic makeup and reproductive trends, a regenerative agenda seeking to rebalance the negative impact perceived to result from the increased proliferation of "scientifically" evidenced "inferior" portions of its human capital. Eugenics thus embodies an authoritarian political strategy for the state's intervention to improve the nation's hereditary substance by discouraging the inferior or flawed from reproducing ("negative" eugenics), while encouraging those deemed genetically superior ("positive" eugenics). In other words, eugenics implies a three-pronged approach: first, defining the subject and recasting the nation in a distinctly biological framework; second, the application of these biological parameters towards assessing and qualifying the perceived degree and agents of degeneration; and third, the implementation of population policies geared towards the alleviation and inversion of this degenerative process.

The definition of fascism employed here foots upon Roger Griffin's "new consensus" approach to defining fascism's cultural and national ambitions, due to its utility in explaining how and why a generic ideal type crystallized as structurally distinct but substantively linked permutations across Europe and beyond. Thus defined, fascism it is a more readily useable evaluative tool with which to dissect the Saxon Self-Help's ideological core, removed from the customary straightjacket of German National Socialism. The Self-Help is hence understood here as a distinctly Saxon permutation of the "palingenetic rebirth myth" definition of generic fascism, which Roger Griffin first introduced in his 1991 *The Nature of Fascism*, and which he revisited in his seminal *Modernism and Fascism*:

> Fascism is a revolutionary species of political modernism in the early twentieth century whose mission is to combat the allegedly degenerative forces of

> contemporary history (decadence) by bringing about an alternative modernity and temporality (a "new order" and a "new era") based on the rebirth, or palingenesis, of the nation. Fascists conceive the nation as an organism shaped by historic, cultural, and in some cases, ethnic and hereditary factors, a mythic construct incompatible with liberal, conservative, and communist theories of society.[15]

In short, fascism's conceptual core is defined as a modernist political religion/ideology (rather than a simplistic daydream of past glories), rebelling against the perceivably decadent incarnation of modernity that it experiences; a revolution whose mystical voyage is motivated by the subjective need to redefine and re-enchant a reinvented "sacred canopy," to replace the sense of meaning and belonging buried under a myriad of distinctively degenerative impulses ascribed to an existing modernity in an all-encompassing act of "creative destruction." Indeed, "modernism can be seen as an attempted rebellion against Modernity carried out in order to inaugurate a new modernity."[16] And yet, Griffin's long-standing ambitions for a "new consensus" bridging the classic East/West divide and coalescing the remarkably broad spectrum of definitions is far from fulfilled, and the conceptual maze that "fascism" has become is perhaps nowhere illustrated as profitably as in his 2006 *Streitforum* on *Fascism Past and Present*, coedited with Werner Loh and Andreas Umland.[17] In his programmatic lead article on "Fascism's New Faces (and Facelessness) in the 'Post-Fascist' Epoch," Griffin offers a condensed rendition of his one-sentence definition, namely that "fascism is a political ideology whose mythic core in its various permutations is a palingenetic form of populist ultra-nationalism."[18] This provides the baseline for a fascinating discussion involving dozens of fascist studies' luminaries, including, amongst others, contributions by David Baker, Jeffrey M. Bale, Martin Durham, Roger Eatwell, Peter Fritzsche, A. James Gregor, Aristotle A. Kallis, Ernst Nolte, Kevin Passmore, Stanley G. Payne, David D. Roberts, Andreas Umland, and Wolfgang Wippermann.

---

15 Griffin, *Modernism and Fascism*, 181.

16 Ibid., 53.

17 Griffin, Loh, and Umland, eds., *Fascism Past and Present, West and East*.

18 Griffin, "Fascism's New Faces," 41.

While the critiques offered range from tacit, apprehensive consent to outright dismissal, a recurrent factor concerns the tension between purely conceptual ideal types and actual, empirically grounded case studies.[19] This dichotomy is usefully navigated by Aristotle Kallis and António Costa Pinto's edited volume on *Rethinking Fascism and Dictatorship in Europe*, featuring insightful chapters by Roger Griffin, Roger Eatwell, and Constantin Iordachi, amongst others. As the editors explain in relation to the process of "dynamic hybridization"—that is, the process of discourses influencing different settings that in turn "recontextualize" them—it "suggests a methodological framework that recognizes the heuristic value of generic phenomena but at the same time seeks to trace their historical trajectories by exploring how individual actors perceived and operationalized them at different stages and in different contexts."[20] A core challenge in interpreting fascist ideologies through the prism of ideal types is hence how to apply a theoretical, "ideal" yardstick to actual case studies in such a way that it can remain elastic enough to account for variation, change, and contradictions; but not be so broadly staked that it becomes a meaningless abstraction removed from actual historical realities.

What has firmly emerged from the decades of concerted study, though, is an appreciation of fascism as comprising an actual ideological or cultural program rather than simply being an anti-movement, further accentuated with Stanley Payne's construct of a matrix that could also account for change over time (as, for example, some movements migrating across the political spectrum). George Mosse and Emilio Gentile, in particular, came to emphasize the emphatic aspect of fascist ideology, and Gentile's work on the aestheticization and sacralization of politics can very profitably account for the propensity of Wilhelm Schunn's national neighborhood to construct its own myths and rituals, saturated with religious language and stylized relics.[21]

So what makes the Saxon example so interesting? For one, a sense of self in flux. The Transylvanian Saxons are in many ways political "inbetweeners," having historically inhabited the grey area between ethnic minority and self-government (if not statehood). Immigrating to Transylvania in

19 For a quick view of the range of case studies potentially included under the fascist umbrella, see Blamires and Jackson, *World Fascism*.

20 Kallis, *Rethinking Fascism*, 5.

21 Gentile, "Fascism as Political Religion."

the latter half of the twelfth century, the *Adreanische Freibrief* that awarded the Saxons a significant degree of autonomy in the thirteenth century also laid the foundation for a distinctly legalistic sense of identity—one framed by the plethora of contracts and decrees that had allowed them to retain a certain sense of independence.[22] This was the case until the Austro-Hungarian *Ausgleich* of 1867, when the Saxons found themselves "denigrated" to the status of an ethnic minority, and their self-perception as a civilizing force in the region was itself subjected to significant assimilatory pressures. This fundamental change in geopolitical status was to lay the foundation for the quest to find a new sense of national meaning, of mission, of being—a process that culminated in an increasingly radical eugenic discourse, and the complementary fascist movement that subsumed it in the late 1920s. This is where the prefix "re" in nation (re)building becomes relevant—it signifies the gradual process of recasting Saxon identity, namely its transition from linguistic-cultural paradigms to an overtly biological understanding of nationhood. In other words, the Saxon eugenic discourse began as a response to, and offered the definition of a perception of degeneration as evidenced over time. It is this distinct ability to accurately trace historic trends that confers the Saxon eugenic discourse the political and cultural depth to complement its ideological breadth.

A second reason for taking a closer look at the Saxons is given by one of the Self-Help's key ideological components, pan-Germanism. Suffused with fascist dogma by the late 1930s, the indigenous Saxon eugenic discourse offers an enticing comparative framework through which to expand the debate to include and juxtapose with those programmes advanced by other minorities in Romania, and indeed Eastern Europe more widely (while we already know, for example, that the neighborhood model of eugenic honorary gifts became an export within Romania, and especially with regards to the Baltics). The analysis of Saxon eugenics offered here is to be understood as both a case study and a stepping stone, an opportunity to compare and contrast it with eugenics potentially advanced by other ethnic minorities, and to thereby rethink the relationship between eugenics and ethnic

22 The importance of legal documents as pillars of Saxon identity is reflected by the title of Heinrich Siegmund's 1931 apocalyptic book on Saxon degeneration *Twilight of the Germans in Transylvania,* an allusion to the God Wotan, bound and gagged by the very laws and contracts that had raised him to power. See Siegmund, *Deutschen-Dämmerung in Siebenbürgen.*

minorities more widely. Therefore, the aim is to augment historiography's perception of eugenics as a state-wielded tool of victimization and assimilation with another perspective, namely that of how and why a biological understanding of identity was ideally suited to an ethnic minority striving towards empowerment and re-homogenization, towards a retrenched "eugenic fortress" in an ethnic wilderness.

While the Saxon historical experience within interwar Romania has already been subjected to considerable scrutiny and written about widely, the ideas, visions, and policies pursued by an increasingly interventionist eugenic dogma have not. However, this forgotten episode has more to offer than a tantalizing case study. Excavating the Saxon eugenic discourse has also yielded a remarkable amount of currently untapped empirical demographic and socioeconomic data, a vast statistical treasure trove of quantitative data invaluable to any study of Saxon, and indeed—due to the questionnaires' frequently astonishing level of detail—any other survey of the wider socioeconomic trends in interwar Transylvania. That said, a critical comparison of these sources with their Romanian counterparts would offer a valuable new research avenue into the omnipresent struggle to define and refine the symbolic geographies and socioeconomic realities of interwar Transylvania's ethnic mosaic.

The singularly most fascinating aspect of the Saxon experiment with national renewal is also the most obvious—the very fact that it produced both a cohesive and clearly elaborated eugenic discourse, as well as an indigenous fascist movement for renewal that amplified, politicized, and eventually institutionalized its vision of national salvation in the 1930s. Fritz Fabritius's Self-Help was seminal in both sculpting Saxon eugenics and translating its tenets into practice, and offers itself to fascist studies as a remarkable new case study of how and why an ethnic minority embarked upon a eugenically ordained quest to re-enchant space and time.

That said, what did Saxon eugenics want, how did its conceptual and methodological core chime with the plethora of diverse eugenic movements that litter interwar historiography and beyond? What about it was distinct, indicative of local needs and means, and what aspects resulted from a process of appropriation and adaption? More concretely, how are the overarching theoretical constructs of eugenics and fascism employed here defined in the first place, and how do they link the Saxon case to the wider, international dimension?

Saxon eugenics' perception of an acute existential crisis gravitated around two overarching but intricately related themes, namely the ongoing loss of *Lebensraum* on the one hand, and a continuously slumping birthrate that was robbing it of its human capital on the other. The contentious issue here was not the fact that the Saxons had been diminished numerically and territorially in the past, but the issue of agency. Did a diminishing *Lebensraum,* as Siegmund would have it, force a demographic decline, or rather, as Csallner and most other Saxon eugenicists vehemently argued, did a declining birthrate allow the ethnic "others" allegedly streaming from the hills a foothold? The implications are obvious in that Siegmund vigorously maintained that the Saxons were actually overpopulated and needed to expand and recolonize lost ground to alleviate their *Raumnot* (therefore, his Land Preservation Society), while Csallner troubled himself with how to move towards an overpopulation in the first place (through his Society of Child Enthusiasts). This fundamental difference unavoidably led to the mentor and student falling out and publicly criticizing each other's work in the 1930s, but both were unequivocal in demanding the expulsion of the foreign "invaders."

Aggravating the catastrophic proliferation of a single- and two-child system, Saxon eugenicists were particularly concerned to find their "sociobiological" studies on differential fertility prove that the nation's most valuable socioeconomic castes (namely priests and teachers), in particular, were teetering on the cliff of an abyss. In other words, there were three core dysgenic processes at work: the loss of *Lebensraum,* the loss of numeric strength, and the loss of hereditary quality as a result of differential fertility.

These issues were, in turn, amplified by a host of tributary concerns, especially mixed marriages, the emigration of racially valuable stock, the employment of non-Saxon agricultural laborers and apprentices, and social ills such as alcohol and tobacco. Saxon eugenicists repeatedly pointed to the substantial cumulative wages Saxons paid the very ethnic others who proceeded to buy their way into Saxon towns and settlements. If this marked a perpetual downward spiral, then the Saxon consumption of alcohol and tobacco only amplified the problem. Consequently, recurrent themes relate to the amount of national wealth wasted on societal scourges, and how much land, how many farms, etc., could have been saved had these funds been deployed in the service of the eugenic cause rather than squandered.

Saxon eugenics in general, and Siegmund in particular, were keenly in tune with developments in Germany, and many Saxon organizations took inspiration from their German counterparts. For example, Csallner's Society of Child Enthusiasts looked to the Reichsbund der Kinderreichen, and Fritz Fabritius's Self-Help had been informed by Heinrich Dolle's economic principles outlined in his *Aus Not zu Brot.*[23] Siegmund, for his part, had extensive links to numerous German and Austrian organizations, the correspondence with them spanning thousands of pages of his personal files held by the Romanian National Archive in Sibiu.[24] A valuable example of the extent to which Saxon eugenics directly benefited from German material and financial help in the 1930s is racial anthropology, and the various studies conducted at Eugen Fischer's behest (as discussed in the following chapter). The same is true of the level of funding awarded to individual eugenicists in the form of research grants or even salaries (as was the case with Csallner and Bredt), most commonly from the Deutsches Ausland-Institut in Stuttgart. Apart from these extensive dealings with German and, interestingly, only to a lesser degree Austrian organizations, Saxon eugenics was notably insular, and there is little evidence that the British, American, French, let alone Hungarian or Romanian movements were discussed at length.

II. Exclusions

As with any such study, this investigation of Saxon eugenics unavoidably demanded incessant pruning. However, as every branch that regrettably had to fall by the wayside of this exploration is itself worthy of an independent investigation, the following selection from the long list of exclusions will hopefully be understood as a mapping of future opportunities for research rather than missed opportunities.

Perhaps the singularly most perplexing exclusion, the Romanian state does not feature prominently in the research presented here. There are two overarching reasons for this. First, and most importantly, the Saxon eugenic and fascist discourses were utterly disengaged from their Romanian coun-

23 Although Dolle vanishes from the Self-Help's discourse rather quickly, his book's importance to early ideology is also acknowledged in Schunn, *Weg und Feinde der NEDR*, 11–12.

24 NA Sibiu, Brukenthal Inv. 107, LL. 1–29, files 38–107.

terparts (with the remarkable exception, of course, of the unpopular and short-lived "friendship pact" between Fabritius and Cuza in 1932).[25] There was relatively little contact, let alone a substantive transfer of ideas, between the two. In fact, the Self-Help commonly referred to its dealings with the state as "foreign policy." As will be argued throughout, the Saxon discourses barely busied themselves with developments outside of their direct spheres of influence (apart from extensive coverage of events in Germany and Austria), thus displaying a remarkable degree of self-absorption.

Second, and differentiating between the Romanian state and the Romanian nation, Saxon eugenics and fascism alike tended to perceive the state as something of a malevolent force of nature willfully assaulting Saxon economic and political viability, from which the Saxon politicians, it was increasingly believed, had failed to shield the Saxon nation. There can be no doubt, of course, that the Romanian state's agricultural and bureaucratic reforms inflicted substantial damage on the Saxon political status quo, economy, and church—and thereby fertilized the ground in which the Self-Help in particular would grow—but the state was clearly the lesser of the two evils compared to the Romanian nation: the organic mass that, according to the rhetoric, swept across Saxon lands robbing them of their *Lebensraum*, their jobs, and their hereditary virtues. Hence, the eugenic discourses were more concerned with the Romanian nation, with the individual Romanian who ventured to buy a Saxon farm or "take" the place of a Saxon apprentice, rather than with the Romanian state's policies. The eugenic fortress was predominantly designed to protect Saxons against the host nation, not the host state.

This peculiar stance towards the Romanian state is perhaps nowhere as apparent as in the eugenic and fascist movements' overtly pan-Germanic ideals and visions of a centralized federation of Romania's Germans. In other words, the Self-Help's loyalty was less towards the Romanian state than to its borders. Greater Romania acquired a large and heterogeneous assortment of new ethnic minorities in the wake of the First World War, none of which are covered here. Due to their remarkably varied historic, sociocultural, economic, and political trajectories (not to mention size), their inclusion would have necessitated a comparative framework the com-

25 For an overview, see Roth, *Politische Strukturen*, 180–82; and Glass, *Zerbrochene Nachbarschaft*, 351–55. See also the surviving correspondence between Fabritius and Cuza, NA Sibiu, CGR, A.I.4, pp. 1–44.

plexity of which this study simply cannot accommodate.

One of the most regrettable omissions is that of gender relations, all the more so as the few broader studies of interwar Saxon women's movements and organizations do not discuss their views on and interaction with the eugenic discourse explicitly. Hence, there remain significant gaps in our knowledge of the roles assumed by the various political and church-based women's organizations that established themselves in the 1880s.[26] There are, however, a range of emblematic debates on the role of women in Saxon society and politics throughout the first half of the twentieth century that invoked race-hygienic themes above and beyond the insistent motherhood imperative of the interwar period, such as with regards to women's education and their struggle for suffrage within the Saxon political structures. For instance, in the pages of the journal *Karpathen* we find an illustrative exchange between Karl Jickeli, who insisted racial hygiene demanded wives and mothers,[27] and Meta Römer-Neubner, insisting on emancipation and the realities of an era in which social norms were being rewritten.[28] Unsurprisingly, Siegmund also got embroiled in these debates, publishing "Sächsische Volksmehrung und Frauenbewegung" ("Saxon Population Growth and the Women's Movement") as a two-part article.[29]

While more extensive research in these areas, as well as biographic studies on the main figures such as Adele Zay, Grete Teutsch, and Lotte Binder will doubtlessly enrich our appreciation of women's involvement with interwar Saxon eugenics, the overall tone assumed by the German-Saxon Women's Union (Deutsch-Sächsische Frauenbund) is clearly heard in its 1934 *Instructions for the Saxon Women (Merkblatt für die sächsische Frau)*. Reflecting a number of the eugenic themes discussed here, two (of a total of sixteen) recommendations on mixed marriages and *Lebensraum* deserve a special mention: "2. Only marry a man of your blood and religion! You will otherwise, from the beginning, plant discord into your home"; and "6. Hold on to your house and land! By selling it you actually take away the soil wherein

26 For a brief overview, see Gabel, "'Wir wollen frei sein, um zu dienen.'" For an interesting collection of oral histories, see Liebig, "Siebenbürger Sächsinnen zur Zeit des Nationalsozialismus."

27 Jickeli, "Mädchenerziehung und Rassenhygiene."

28 Römer-Neubner, "Frauenbewegung und Rassenhygiene."

29 Siegmund, "Sächsische Volksmehrung und Frauenbewegung, Teil I und II." Also see the ringing endorsement from Grete Teutsch in her subsequent article, "Dr. Heinrich Siegmund und die Frauenbewegung."

your descendants' strength is rooted!"[30]

If nothing else, it is hoped that this foray into Saxon eugenics will convey, first, the core ideological leitmotifs and complementary sense of an existential crisis that molded and guided this budding eugenic discourse. Second, it aims to illustrate the extent to which eugenics and racial hygiene were perceived as offering an embattled ethnic minority a novel methodology with which to redefine its national sense of self, to diagnose the "cancer" afflicting its organic national body, and to formulate a scientifically quantifiable and verifiable "cure." Third, it will explore the manner in which these ideological trends are illustrated by Alfred Csallner's embrace of a eugenic discourse that could translate his sense of a statistically evidenced impending national extinction into a viable program of national salvation. It is hoped that this exploration will yield useful insights into the mechanisms of how and why an ethnic minority in general, and a priest in particular, turned to eugenics—and ultimately fascism—in an increasingly desperate attempt to avert a despairing battle for national renewal both internally, and vis-à-vis its Romanian host nation.

III. Unpacking the Past

The study of interwar eugenics has enjoyed a veritable renaissance over the past three decades, one that has recently propelled the discipline's diversification beyond the "classic" case studies offered by the United States, Great Britain, and an omnipresent National Socialist Germany that dominated the discourse in the 1980s and 1990s.[31] Standing on the shoulders of these pioneering studies, the investigation of hitherto neglected or forgotten eugenic movements—as diverse ideologically and methodologically as they are spread geographically—has undoubtedly enriched our understanding of one of the twentieth century's most consequential ideologies.[32] This

30 Deutsch-Sächische Frauenbund, ed., *Merkblatt für die sächsische Frau*, n.p. For a full translation, see Georgescu, "Transylvanian Saxon Eugenics."

31 See, for example, Weindling, *Health, Race, and German Politics*; Weindling, *Nazi Medicine and the Nuremberg Trials*; Weiss, *Race, Hygiene, and National Efficiency*; Burleigh and Wippermann, *The Racial State*; Proctor, *Racial Hygiene*; Schmuhl, *The Kaiser-Wilhelm-Institute for Anthropology, Human Heredity, and Eugenics*; and Blackbourn, *The Conquest of Nature*.

32 Amongst the plethora of studies that this wider geographic focus has produced, see Klausen, *Race, Maternity, and Politics of Birth Control in South Africa*; Dikotter, *Imperfect Conceptions*; Eraso, "Biotypology, Endocrinology, and Sterilization"; Stepan, *The Hour of Eugenics*; Cleminson, *Catholicism, Race, and Empire*; as well as

remarkable conceptual and empirical expansion is in many ways an accidental reflection of the manner in which a global audience had appropriated and adapted a eugenic discourse initially born of West-European fears of a degenerative, industrialized modernity—a conceptual stream that quickly turned into a torrent as it meandered across the continent and swelled with the emergence of ever more tributaries.[33]

This is a particularly exciting time to be studying interwar eugenics. Historiography has recently begun not only to take a closer look at the "negative" eugenic policies wielded against those segments of a population designated "dysgenic," but also to explore how and why a biological/organic definition of nationhood captured the public and political imagination alike at a time of existential crisis, real or perceived. As the blank pages in the history of eugenics are gradually filled, Eastern Europe has increasingly become a focus of academic scrutiny. This development was most appropriately signaled by Maria Bucur's pioneering 2002 book on *Eugenics and Modernization in Interwar Romania*, and further expounded in Marius Turda and Paul Weindling's landmark 2007 edited volume on *"Blood and Homeland": Eugenics and Racial Nationalism in Central and Southeast Europe*, and subsequently in the 2011 volume edited with Christian Promitzer and Sevasti Trubeta on *Health, Hygiene, and Eugenics in Southeastern Europe to 1945*. The most comprehensive study, both in terms of reproducing sources and its breadth of analysis, has just been published, namely Marius Turda's 2015 edited volume on *The History of Eugenics in East-Central Europe, 1900–1945*.

Maria Bucur's *Eugenics and Modernization in Interwar Romania* was the first to scrutinize Romanian eugenics and, perhaps most importantly, to focus not merely on the movement's key figures, but on the role eugenics played in shaping national identity. This eruption in scholarly interest in Romanian eugenics specifically, and the East European history of med-

the fantastically broad geographically and thematically Bashford and Levine, eds., *The Oxford Handbook of the History of Eugenics*.

33 For example, the German Society for Racial Hygiene was founded in 1905, the British Eugenics Education Society in 1907, further eugenic societies were founded in Prague and Vienna in 1913, while the Hungarian Society for Racial Hygiene and Population Policy and the Polish Society for the Struggle against Race Degeneration convened in 1917. Romania produced several eugenic organizations: Iuliu Moldovan established the Eugenics and Biopolitical Section of the Astra Association in Transylvania in 1927, Sabin Manuilă created the Demographic, Anthropological, and Eugenics Section of the Romanian Social Institute in 1935, and the same year the Romanian Royal Society of Eugenics and Heredity was founded under Gheorghe Marinescu's presidency.

icine more widely, has undoubtedly benefited from the pioneering studies produced by Marius Turda, whose publications have not only served to excavate a forgotten case study, but have also related the Romanian and Hungarian experiences and experimentation with eugenic and biopolitical thinking to the wider, Pan-European discourses that informed them.[34]

This exponential growth in academic output was facilitated by a conceptual reevaluation of interwar eugenics as not merely a permanent stain on European medical ethics, but a useful reflection of academic and popular currents in interwar intellectual history, and of the diverse nation (re)building projects of its time. It is precisely this sociocultural dimension—i.e., the eugenic *Weltanschauung*'s ability to simplify and condense complex historical processes into easily recognizable biological terms predicting a nation's future with the authority of a white lab coat and mathematical formulae—which ensured the popular currency of eugenic projects. It is hence little surprising that most of the Saxon eugenicists were neither scientists nor medical professionals—in fact, several of the most important advocates of eugenics were priests and teachers who utilized the eugenic vocabulary to cast perceptions of degeneration in a biological language as unequivocal and prescriptive as it was "modern." It is this realization that justifies the proposition that any venture to understand the underlying causes and consequences of an interwar world in turmoil must necessarily also explore eugenics.

While the Transylvanian Saxon eugenic discourse may have eluded academic scrutiny, the Saxons themselves do occasionally feature in the ever-expanding literature on the Third Reich. When they do, it is mostly either in relation to National Socialist territorial ambitions (and plans for their ethnic reconfiguration),[35] or in terms of their utility as diplomatic bargaining chips vis-à-vis the Romanian state.[36] These studies, however, frequently focus on Germany's appropriation or instrumentalization of Romania's Germans,

34 See the following contributions in particular: Turda, "The Nation as Object"; Turda, "'To End the Degeneration of a Nation'"; Tuda, "In Pursuit of Greater Hungary"; Turda, *Modernism and Eugenics*; Turda and Gillette, *Latin Eugenics in Comparative Perspective*; and Turda, ed., *The History of Eugenics in East-Central Europe.*

35 See, for example, Burleigh, *Germany Turns Eastwards*; Lumans, *Himmler's Auxiliaries*; Ritter, *Das Deutsche Auslandsinstitut in Stuttgart*; Beer and Seewann, eds., *Südostforschung im Schatten des Dritten Reiches*; Fahlbusch, *Wo der deutsche… ist, ist Deutschland!*; and Fahlbusch, *Wissenschaft im Dienst der nationalsozialistischen Politik?*

36 See Haynes, *Romanian Policy towards Germany*; and Haynes, "Germany and the Establishment of the Romanian National Legionary State." Also, Deletant, *Hitler's Forgotten Ally*; and Hitchens, *Rumania, 1866–1947.*

increasingly harnessed to the Reich's biopolitical dystopia after 1940. Posing as pockets of untapped human capital, these German ethnic enclaves were also the focus of intense scientific curiosity, and considerable funding was made available for their anthropological and biometric study as idealized vestiges of a rural and grounded past uprooted by industrialization and the onslaught of a degenerating modernity.[37] The flipside was—at least from the point of view of an isolationist ethnic minority whose "racial purity" went largely unquestioned—that it also amounted to a vindication of a perpetually reiterated "historic mission" to civilize and cultivate, to act as geopolitical and biopolitical bridgeheads. While current interpretations tend to focus on how and why a plethora of National Socialist state and party organizations vied for supremacy over Europe's "*Volksdeutsche*," they often do so by looking in from the outside.

Romanian historiography would be ideally placed to make a valuable contribution to Saxon studies and, in the process, recognize and reconcile itself to the biopolitical, fascist, and communist legacies of the twentieth century.[38] Indeed, the overwhelming bulk of literature produced prior to the collapse of communism is largely omitted from this study, if not for conceptual, then certainly for empirical reasons. As Vasile Ciobanu put it in his particularly valuable 2001 *Contributions to the Understanding of Transylvanian Saxon History*, Romanian researchers' failure to access either the rich source base held by the Romanian National Archives or their German counterparts' work had resulted "in a number of inexact and erroneous interpretations relating to the Saxons sneaking into a series of monographs on the interwar period and Second World War."[39]

What conclusions has Saxon historiography derived from its six-decade introspection? German historiography on the Saxons has largely been dominated by Saxons, and is subject to the inherent difficulties that this situation entails. Whereas this may go a long way in explaining its at times highly

37 A seminal case study on the Banat town of Marienfeld is offered by Maria Teschler-Nicola's magisterial "*Volksdeutsche* and Racial Anthropology in Interwar Vienna."

38 Rebecca Haynes provides a valuable account of the various "myth-building" exercises which dominated the debates on Romanian national identity, arguing that the increased access to Romanian archival material since 1989 must necessarily inform these and other historiographic debates. See Haynes, introduction to *Occasional Papers in Romanian Studies*, No. 2.

39 Ciobanu, *Contribuții la cunoșterea istoriei sașilor Transilvanei*, 21. Also see his 1980 assessment of the source base, Ciobanu, "Neue Quellen zur Geschichte der Siebenbürger Sachsen"; and Ciobanu, *Identitatea culturală a germanilor din România în perioada interbelică*.

polarized and politicized nature, it also points to the comparatively delicate nature of a field slowly slipping from memory into history. While the genesis and proliferation of Saxon eugenics has seemingly gone unnoticed, the hallmark monographs by Johann Böhm, Karl Reinerth, Wolfgang Miege, and Harald Roth are regularly depicted as either too militant, too revisionist, too limited in scope, too independent from available primary sources, or too saturated with them at the expense of analysis. As Harald Roth aptly summarizes: "Miege's view is largely that of the Foreign Office,[40] Reinerth seems incapable of distancing himself from a personal affinity with the Transylvanian Saxons,[41] and Böhm's prime intention is to find and prove that all dominant political developments are related to National Socialism and thus negate any historical independence."[42] Roth's criticism of Böhm's work is particularly relevant in terms of highlighting a common straightjacket—the focus on whether one or the other movement was specifically National Socialist, a contentious issue which has also resulted in rather unfortunate exchanges between the various viewpoints.[43]

Roth's 1994 key monograph *Political Structures and Currents amongst the Transylvanian Saxons, 1919–1933*,[44] is a highly informative analysis of how and why National Socialism established itself as a political force in Transylvania. Roth sets out to investigate when the Self-Help movement became

40 Miege, *Das Dritte Reich und die Deutsche Volksgruppe in Rumänien*. Although Miege's 1972 book inevitably labors under the constraints of its time, he is actually one of the few historians to understand the Self-Help as an indigenous Saxon movement likely to have emerged regardless of political developments in Germany. However, Miege regrettably discounts eugenics and racial hygiene as superficially adapted National Socialist imports.

41 Roth's judgement is rather lenient. Founded in 1965, Karl Reinerth's Work Group for Southeast European National and Homeland Studies (Arbeitsgemeinschaft für südosteuropäische Volks- und Heimatsforschung) attracted a number of Saxon fascists, such as Wilhelm Staedel, Misch Orend, and Alfred Bonfert. See Reinerth and Cloos, *Zur Geschichte der Deutschen in Rumänien*; Reinerth, *Zur politischen Entwicklung der Deutschen in Rumänien*; and Reinerth, "Zu den Innenpolitischen Auseinandersetzungen." In fairness, though, and as Reinerth states throughout the latter volume in particular, his is an overtly subjective approach to Saxon interwar politics.

42 Roth, *Politische Strukturen*, 14.

43 A recent example revolved around Paul Milata's assessment of the involvement of Romania's Germans with the Waffen-SS: Milata, *Zwischen Hitler, Stalin und Antonescu*. Klaus Popa led the charge on what he deemed "neoconservative propaganda," dismissing Milata's book as "an experiment in presumptuousness, prejudice, and bias, prefabricated explanations, sweeping judgements, self-righteousness, self-referentiality, and self-pity." Popa, "Neokonservative Propadandaschrift," 14. A polarized exchange between Milata and Böhm subsequently erupted over who "plagiarized" Andreas Schmidt's 1944 CV from whom. See Milata, "Anmerkungen zu einem Plagiatsverdacht"; and Böhm, "Anmerkungen zu einem Plagiatsverdacht."

44 Roth, *Politische Strukturen*. In addition, Roth has edited a wide range of books including, amongst others: Roth, ed., *Minderheit und Nationalstaat*; and Roth and Hausleitner, eds., *Der Einfluss von Faschismus*.

politicized, and when it officially adopted a specifically National Socialist ideology—both of which he locates in the early 1930s.[45] Given that Böhm insists this adoption of German National Socialism had in fact been evident as early as 1922, it is little surprising to find Roth dismiss him as both conceptually flawed and methodologically unsound.[46]

Roth's conclusions on Saxon interwar political processes and the manner in which they might have allowed for their gradual infiltration by National Socialist ideology are highly indicative of the overall consensus in Saxon historiography. Broadly, Roth outlines a Saxon political establishment suspicious of dissent within its ranks—an "obedience-driven democracy" (*Gefolgschaftsdemokratie*) that chastised opposition as treacherous and undesirable, as an inherent weakening of the collective's cohesion and bargaining power vis-à-vis the state.[47]

Turning to the question of how the Self-Help managed to appeal to the wider population, Roth refers to a set of joint values thought to have rendered the Saxons susceptible to National Socialist lures:

> As essential elements of this ideology had precedents in Saxon traditions, Fabritius could easily propagate the movement at a time of particular insecurity and dissatisfaction, and his support base grew quickly. Ethnocentricity, a pronounced sense of community, mutual social aid, and a traditionally cooperative approach to the economy were established tenets of the Saxons' self-perception. And these were easily transferred unto a National Socialist ideology. The National Socialist race ideology, as a tool towards self-preservation in a multiethnic environment, was similarly comprehensible to its contemporaries; the previously nonexistent anti-Semitism had to be imported and proliferated first.[48]

There are various points of interest in this excerpt. Starting with the somewhat awkward concept of anti-Semitism as an "import," this passage

45 Roth, *Politische Strukturen*, 177, 219.

46 Unreservedly blunt, Roth has the following to say about Böhm's *Die Deutschen in Rumänien und die Weimarer Republik*: "He largely restricts himself to overall discussions, though, does not apply the proper methodological tools [*Quellenkritik*], and has the apparent goal of locating National Socialism amongst the Germans in Romania as early as possible. [. . .] This work must be understood as an intentional historiography [*intentionalen Geschichtsschreibung*], and hence considered an unsuccessful attempt to portray the years 1919–1933." Roth, *Politische Strukturen*, 12–13.

47 Ibid., 219.

48 Ibid., 217–18.

presumes a grafting of National Socialist ideas onto generalized societal norms, implying passivity somewhat removed from the sense of crisis and revolutionary zeal so characteristic of the Self-Help.

Diametrically opposed to this line of reasoning, one finds Johann Böhm.[49] Having written six monographs on the interwar period, he is certainly prolific, if controversial. Based on an impressive body of empirical research, Böhm relentlessly argues that the Saxon interwar discourse had been saturated with (rather than usurped by) explicitly National Socialist ideology as early as 1922. Böhm has been particularly vocal in his demand that the Saxons "come clean" about their National Socialist past. His *magnum opus* on the Saxon interwar experience began with his 1985 doctoral dissertation, and concluded in 2003 with the final volume of his published trilogy. While an essential guide to the period 1919–44, Böhm's conclusions are in many ways similar to Roth's, especially in pointing to the unreceptive Saxon political status quo as the crucial platform upon which the Self-Help could build its agenda and garner public support. Böhm also insists that although Saxon fascism had begun with Fritz Fabritius in 1922, its eventual empowerment was a consequence of the National Socialist rise to power in Germany, coupled with the Romanian state's curtailing of Saxon political rights.[50]

Böhm's ambition also suffuses his latest book, *Hitler's Vassals in the German Community in Romania before and after 1945*. Grounded in an impressive array of archival material, Böhm provides biographies for fifteen of the most prominent Saxon journalists, authors, priests, and officials. Although Böhm has begun plugging one of the gaping holes in Saxon historiography, especially as several of the biographies are exceptionally useful, this book is unrelenting in its accusations of historical revisionism.[51]

One book to have impressively transcended this historiographic bottleneck is Hildrun Glass's *Broken Neighborliness: The German-Jewish Relationship in Romania, 1918–1938*.[52] The 1995 study is an outstanding, over

49 Böhm, *Das Nationalsozialistische Deutschland*; Böhm, *Die Deutschen in Rumänien und die Weimarer Republik*; Böhm and Braeg, eds., *Dr. Viktor Glondys*; Böhm, *Die Deutschen in Rumänien und das Dritte Reich*; and Böhm, *Die Gleichschaltung*. For a more concise overview, see Böhm, "Die politische Entwicklung"; and more recently, Böhm, "Von der völkischen Verblendung."

50 See Böhm, *Das Nationalsozialistische Deutschland*, 24, 212.

51 "A large proportion of the Germans in Romania became National Socialists [...]. That they deny their responsibility following the defeat of 1944/45 and acted as if they had been only innocent victims, is to ridicule historical facts." Böhm, *Hitlers Vasallen*, 1.

52 Glass, *Zerbrochene Nachbarschaft*. See also her article on Saxon anti-Semitism and its impact on individ-

600-page-long analysis of German-Jewish relations in interwar Romania. Making use of a broad spectrum of primary sources, this book covers not only the Saxons, but also the Swabians of the Banat and Bessarabia, and the Germans in Bukovina. *Zerbrochene Nachbarschaft* is also the only book to locate the eugenic within the Self-Help's discourse (which she insists had advocated a political platform from the outset), albeit in passing and exclusively in terms of its anti-Semitic content.[53] Glass's book is an essential reading for any scholar active in this field, and the singularly most insightful analysis of Saxon interwar anti-Semitism.

In sum, Saxon historiography tends to focus on perpetual speculations over when the Self-Help assumed a distinctly political and National Socialist guise, an undoubtedly crucial question. Nonetheless, the existing body of work all too often overlooks what ultimately constituted an indigenous Saxon fascism prior to this transition. While the political establishment's inability to accommodate and address dissent is certainly a key factor, the self-perception of a nation in rapid decline is paramount to explaining the rise of a eugenic discourse which preceded the politicized movement for renewal that eventually subsumed it.

ual families (including a brief discussion of Schunn's neighborhoods and their opposition to mixed marriages): Glass, "Wer ist ein Deutscher?"

53 She deflates this claim rather sharply when she writes: "Historians of the German minorities write instead that there had been no [*fremd gewesen*] anti-Semitism amongst the Transylvanian Saxons before 1933; that the National Socialist race ideology had not met with the approval of a majority of the German minorities." Glass, *Zerbrochene Nachbarschaft*, 15.

CHAPTER I

LOCATING AND DEFINING THE TRANSYLVANIAN SAXON EUGENIC DISCOURSE

Who were the key protagonists of the Saxon eugenic discourse as it emerged and evolved into its *Gleichschaltung* in 1940, how did they define the dysgenic crisis, and how did they seek to alleviate it? Saxon eugenics always revolved around the twin themes of *Lebensraum* and a looming demographic collapse, but they were weighted differently by different actors. How do these individuals' ideas and careers reflect the import and impact of local and international networks, from close-knit student associations to prolific Reich institutions that heavily subsidized race-anthropological research? Similarly, how did the population policies they put forward reflect the growing importance of the fascist Self-Help as an alternative route to empowerment compared to the church, and to what extent were they indicative of Saxon fascism's close ties to German racial hygiene?

1. Heinrich Siegmund and the Origins of Saxon Eugenics

In Heinrich Siegmund, Saxon historiography has forgotten one of its most important and exciting protagonists. The handful of biographical sketches do not do justice to Siegmund's significance to Saxon history above and beyond the eugenic discourse he spawned.[1] His vast legacy, while largely

1 See Behrwerth, "Dr. Heinrich Siegmund"; Wagner, "Heinrich Sigmund"; and Linger, "Sitte, Moral und Volksreinheit." Several invaluable biographies were published by Siegmund's close associates after his death, the most insightful being Rehner, *Dem Andenken Dr. Heinrich Siegmunds*. See also the biography and partial bibliography offered by Julius Ernst Gyurgyevich, who took over Siegmund's Welfare Exhibit in 1937, and was also a leading figure in the national neighborhoods: Gyurgyevich, "Vorarbeiten zu einem Schriftenverzeichnis"; and Gyurgyevich, "Aus gesegneter Ernte."

peripheral to the fascist discourse as such, was fundamental to the Self-Help's ideology, so heavily influenced by his friend and student, Father Alfred Csallner. Perhaps more significantly still, Siegmund actively fostered collaboration between his Welfare Office, the Self-Help, and Csallner's Society for Child Enthusiasts in the late 1920s. But Siegmund's influence stretched far wider, and had a formative impact on society at large.

Siegmund produced a flood of publications spanning over forty books, several journals, and literally hundreds of articles unto his death in July 1937. He embarked on a relentless, if largely unsuccessful quest to convince the Saxon establishment of his "displacement theory" (*Verdrängungslehre*), made to account for the Saxon demographic and territorial decline over past centuries. Siegmund insisted the Saxon nation had declined due to internal reasons, namely the failure to actively protect its *Lebensraum*—the ongoing loss of which he thought (with a clearly Malthusian mindset) contributed to the decline in Saxon fertility that, in turn, amounted to an impending "national death" (*Volkstod*) or, as he put it in the title of his 1931 book, a *Twilight of the Germans in Transylvania*.[2] The measures he proposed to revitalize the Saxon settlements, which his "national biology" (*Volksbiologie*) diagnosed as either "endangered," "sick," "seriously ill," or "dying" (*gefährdet, krank, schwerkrank*, or *sterbend*), were subsumed under a "science of national defense" (*Wissenschaft der Volksverteidung*), and formed the bedrock upon which to build a re-homogenized Saxon eugenic fortress. Indicative of the ideological core that was to define Saxon eugenics, Siegmund called for a significant qualitative increase in Saxon fertility; the betterment of the nation's racial stock by discouraging the hereditarily ill and mentally infirm from reproduction; a decrease in emigration; a battle against the corrosive effects of tobacco and alcohol on societal health and the nation's collective purse; and the promotion of internal colonization schemes to create a living belt of racially vibrant settlements enclosing an enlarged and re-homogenized Saxon *Lebensraum*.

2 Siegmund, *Deutschen-Dämmerung in Siebenbürgen*. Although the book received much glowing praise, it also caused some controversy. Karl Kurt Klein criticized its methodology and labelled it as ahistorical in a review he published in the influential *Transylvanian Quarterly*, which he edited. See Klein, "Geschichtswissenschaft und naturwissenschaftliche Pseudohistorie." The journal also published a much more favorable review by Hermann Oberth: Oberth, "Ist Siegmunds 'Deutschen Dämmerung' unwissenschaftlich?" Siegmund, incensed, sent Klein a response for publication, but Klein refused to publish it.

Figure 1. Heinrich Siegmund.

An aptly succinct abstract of Siegmund's wide-ranging catalog of grievances and policies, from language purity to economic autarky and mixed marriages, comes in the form of his 1928 welfare flier detailing "national moral imperatives" (*Völkische Sittengebote*) and insisting that "every nation must always be on guard. Every national comrade should remind himself daily that he, from morning till evening, must like a warrior in the trench fight for his nation's victory in the struggle for survival."[3] Siegmud offered a staggering seventy-nine imperatives grouped into five sections on how to "increase the number of national comrades" (nos. 1–9), "keep our race pure und healthy" (nos. 10–15, praising Csallner's VdK and protesting mixed marriages), "tend to the health of the Saxon national body" (nos. 16–55, including advice on only buying from and hiring Saxons, the imperatives to have at least seven children, marry young, donate to the welfare committee, etc.), "tend to your Germanness" (nos. 56–73, including an interesting

3 Also published as a two-part article, "Völkische Sittengebote: Teil I und II."

denunciation of Esperanto and other "impossible world-languages"[4]), and to "increase the Saxon *Lebensraum*!" (nos. 74–79). Siegmund was the first Saxon to write about racial hygiene for wider audiences, as illustrated by his landmark 1901 article "On Saxon Racial Hygiene," published in the main Saxon newspaper *Siebenbürgisch-Deutsche Tageblatt*. In it, quite remarkably, Siegmund called for the creation of a race-hygienic society, and set out to "illustrate the importance of a health science [*Gesundheitswissenschaft*] informed by the new realizations about life, as applied to the Saxon nation as a race."[5] Elaborating on its wider educational purpose and necessarily diverse or holistic membership base, from economic to medical professionals, Siegmund considered it must be "the society's endeavor to reconcile, as far as possible, the contradiction between individual and racial hygiene, which by its inherent imperative [*Naturnotwendigkeit*] admonishes the conflict between egoism and altruism, selfishness and brotherly love, by carefully investigating and evaluating the corresponding perspectives' expectations according to rich practical and theoretical experience."[6] In promoting a vision of a racial hygiene grounded in historic practices but underpinned by notions of scientific rigor and objectivity, Siegmund offered a popular audience both a cure for the present and a promise for the future: "Should we Saxons manage to follow the laws of racial hygiene based upon the broad and stable ground of scientific experience and insight—and which exhibits so many similar traits to the thousands-year-old, tried and tested commandments of the ancient Near-Eastern civilizations, above all the Jews—then our national future would be secured for quite some time."[7]

Siegmund, who was the grandson of the Saxon national hero Stephan Ludwig Roth, studied medicine in Graz and Vienna, and was heavily influenced by the works of Thomas Malthus, Charles Darwin, Ernst Haeckel, and Rudolf Virchow. Upon his return to Mediasch in 1893, Siegmund opened a

4 Siegmund was incredibly active when it came to language purity, especially with regard to terms he considered foreign and for which he concocted "German" alternatives; this did little to make his texts more accessible, if certainly distinctive. Siegmund even founded the short-lived Society for German Language and Print in Romania (Deutschen Sprach- und Schriftenvereins in Rumänien) in Mediasch in 1919; its local branch in Mediasch wrote to the regional committee (*Kreisauschuss*), requesting that it "Germanizes" thirteen passages of its national program by changing, for example, "Funktion" to "Aufgaben" and "Nation" to "Volk." See NA Sibiu, Brukental Inv. 107, LL.1-29.76, p. 64.

5 Siegmund, *Zur sächsischen Rassenhygiene*, 3. For a longer translation from this text, see Georgescu, "Transylvanian Saxon Eugenics."

6 Siegmund, *Zur sächsischen Rassenhygiene*, 20.

7 Ibid., 24.

medical practice and became the city's medical officer (*Stadtphysikus*) two years later. Siegmund founded the first of his numerous journals in 1902, but as the *National Health* (*Volksgesundheit*) was (and still is) largely ignored, it ceased publication in 1911.

In early November that same year, Siegmund convened a long-anticipated symposium to "discuss Saxon national matters" (*Beratung Sächsischer Volksfragen*) together with Carl Wolff, who also chaired it. While this association between two of the most important Saxon *völkisch* protagonists is interesting in its own right, the symposium also featured lectures by Siegmund on "Forms of National Death" (*Formen des Volkstodes*) and by Heinrich Müller on "Obstacles to Increasing the Number of Saxon Souls and the Issue of Racial Degeneration amongst the Saxons" (*Hindernisse der Vermehrung der sächsischen Seelenzahl und rassische Entartung unter den Sachsen*).[8]

In his opening address, Siegmund elaborated on the need for such a symposium as "it is becoming ever more difficult to defend against the state- and government-promoted attempts at merging [*Einschmelzung*] our nation into the homogenizing Magyar melting pot of nations [*madjarisch-einheitliche Völkervermischung*], and the no less brash international economic distress, caused by the influences of the global economy, but also by the onslaught of Romanians and members of other nations."[9] That said, and following a necessarily emphatic proclamation that the German "nature" was the pinnacle of cultural achievement, Siegmund insisted that "the goal of the scientific ethnology [*Volkskunde*] must hence be the creation of a Saxon racial and social science on the foundation of life's laws [*Lebenslehre*]."[10]

Offering an important example of his extensive links to German eugenicists, it was also in 1911 that he joined Alfred Ploetz's International Society for Racial Hygiene as a founding member, in return for advancing its agenda in Transylvania (the two corresponded as early as 1904).[11] Siegmund sat at the centre of a remarkably extensive network connecting Saxon eugenics with their German counterparts. Apart from prominent individu-

8 NA Sibiu, Brukental Inv. 107, LL. 1-29.84, p. 72.

9 Siegmund, "Eröffnungsrede zur Aussprache über die Lebensbedingungen des sächsischen Volkes und die Mittel der Volksmehrung an der Hand der wissenschaftlichen Erfahrung," January 24, 1911, NA Sibiu, Brukenthal Inv. 107, LL. 1-29.98, p. 330.

10 Ibid., 332.

11 For the correspondence, see NA Sibiu, Brukenthal Inv. 107, LL. 1-29.65.

als such as Alfred Ploetz, Hans Harmsen, and Eugen Fischer, Siegmund also maintained close relations with a range of German and Austrian organizations, such as the German Society for National Improvement and Heredity (Deutscher Bund für Volksaufklärung und Erbkunde),[12] the nationalist and anti-Semitic German Nationalist Protection and Defiance Federation for Austria (Deutschvölkischer Schutz- und Trutzbund für Österreich), and was also a member of the German Association (Deutschenbund), promoting a race-based ideology, since 1917.[13] Of these various personal and professional networks, one of the most interesting is his membership in the Saxon student society Saxonia, in Graz, not only because its members regularly appear as collaborators in Siegmund's various projects, but also because Saxonia joined his Land Preservation Society (Bodenschutzverein, or BSV) in 1917, and donated to its causes.[14]

Siegmund was hence also instrumental to the Saxon search for institutional means to eugenic ends, and initially attempted to translate his agenda through private initiatives, before turning to the church for organizational support in the early 1920s. The BSV's creation in 1906 was a significant moment in the evolution of Saxon eugenics in Transylvania, the first of the host of eugenically minded organizations that followed.[15] Subsuming the various keystone tenets of his "science of national defense," and its insistence on a dwindling *Lebensraum* as the motor driving Saxon degeneration, it set out to employ a "defense fund" (*Wehrkasse*) filled by donations and annual membership fees towards defending Saxon, and acquiring non-Saxon properties, which, in turn, were to be allocated to hereditarily sound settlers.

The creation of the BSV had been decided as early as May 1905, at a public meeting held to discuss the loss of *Lebensraum* in Mediasch, and was followed by an open call for members sent to some 1,000 individuals and organizations in December, but it was not until March 1906 that it was formally

12 Ibid.

13 Ibid., 60–66.

14 NA Sibiu, Brukental Inv. 107, LL. 1-29.79, pp. 7–8.

15 Actually, Siegmund was the BSV's secretary until he decided to become more active again in 1912, replacing R. Schuller, who became his deputy. R. Schuster became the new secretary with Rudolf Brandsch deputizing, while Michael English (who was the chair of the local branch in Leschkirch) became the treasurer, with Heinrich Müller and Franz Liess as deputies. See NA Sibiu, Brukental Inv. 107, LL. 1-29.78, pp. 37, 39.

founded with fifty-five members.[16] In its first burst of activity, and despite repeatedly insisting it had worked secretively until deciding to go public in 1914,[17] the BSV published various brochures, fliers, and lectures, and it even opened an estate agency in Mediasch. With its revised draft statutes eventually approved by the Minister for Agriculture in 1908,[18] the BSV began to recruit more actively, and by the end of 1909 had local branches in Mediasch (406 members), Reps (chair Heinrich Müller, 22 members), and Hermannstadt (chair A. Bell, 125 members).[19] By July 1911, Siegmund was pleased to report to the BSV's executive committee meeting that it had a total of 1,050 members,[20] while the member list for Hermannstadt alone boasted 384 names, including, most remarkably, that of Fritz Fabritius.[21]

Siegmund's brochure on the *Aim and Establishment of the Land Preservation Society* was saturated with his views on heredity and racial hygiene, invoked as objective sciences diagnosing the Saxon national body's decline in terms of the four core variables—"mass, race, national health, and *Lebensraum*."[22] Correspondingly, Siegmund argued that nations suffered four types of "national death" (*Volkstod*): numerically, racially, bodily, and spatially. But it was the perceived ongoing loss of Saxon *Lebensraum* to Romanians that constituted the most acute threat, and "it cannot be doubted, the numbers tell us that much with perfect clarity, that we Saxons appear to be in the inevitable danger of being forever displaced from our villages and cities by the Romanians."[23] As the 1908 flyer *Dear National Comrade!* (*Werter Volksgenosse!*) insisted: "The Romanians penetrate the Saxon lands ever faster! The loss of our land ownership inevitably means the demise of our Saxon nation. The Romanians win one piece of land after the other. That gives them strength for the unstoppable push into the Saxon cities. Even our houses get into their possession!"[24]

16 NA Sibiu, Brukental Inv. 107, LL. 1-29.85, pp. 1–2.

17 NA Sibiu, Brukental Inv. 107, LL. 1-29.78, p. 43. Within this context, it is also interesting to note an exchange between Siegmund and the Verein Südmark, in which the latter somewhat perplexedly informs Siegmund that they had read about the BSV in the "Saxon-American Post" newspaper. See NA Sibiu, Brukental Inv. 107, LL. 1-29.83, p. 40.

18 See NA Sibiu, Brukental Inv. 107, LL. 1-29.76, pp. 1–2.

19 For the annual reports of the three local branches, see NA Sibiu, Brukental Inv. 107, LL. 1-29.85, pp. 3–7.

20 NA Sibiu, Brukental Inv. 107, LL. 1-29.78, p. 30.

21 NA Sibiu, Brukental Inv. 107, LL. 1-29.88, pp. 43–50.

22 Siegmund, *Ziel und Einrichtung des Bodenschutz-Vereins*, 2.

23 Ibid., 7.

24 NA Sibiu, Brukental Inv. 107, LL. 1-29.76, p. 13.

Consequently, the BSV's core aims as found on a flier advertising its *Gratuitous Mediation and Information* (*Unentgeltliche Vermittlungen und Auskunft*) services, asked:

What does the preservation and proliferation of the Saxon nation mean? It means:

1. Preservation and increase of the nation numerically (birthrate excess)
2. Preservation and heightening of the Saxon national comrades' mental and physical aptitude [*Tüchtigkeit*] (racial duty [*Rassedienst*])
3. Improvement and development of the Saxon national body's organization, that is, composition [*Gliederung/Aufbau*]
4. Preservation and ennobling of the Saxon cultural achievement
5. Preservation and increase of the Saxon *Lebensraum*—that means, above all, of Saxon land ownership.[25]

In a 1913 lecture on "What Can and Should Saxon Colonization Work Achieve?," Siegmund lists seven core principles, from demographic to territorial, arguing that: "Only comprehensive and purposeful colonization work can provide in sufficient numbers the indispensable battle troops needed in the economic struggle with the Romanians, namely the agricultural workers." Further, he adds that "as every colonization is a selection process amongst the settlers, and racial imperatives can therein be considered, a comprehensive settlement of Saxon agricultural workers is one of the best measures of Saxon racial care [*Rassenpflege*]."[26] It is then not particularly surprising to find amongst the BSV's terms and conditions for aid for potential colonists two clauses stipulating that abstinent and non-smoking applicants would be preferred, and that "applicants with venereal or mental illness, or afflicted with severe hereditary failures [*Fehlern*] cannot be considered."[27]

Rather interestingly, the BSV launched an essay competition in 1911 on the question "Displacement or Destruction?" that received only a single entry, by a certain Robert Csallner. The jury consisting of Bishop Friedrich Teutsch of GA Schuller decided that the essay had not actually answered the

25 Ibid., 12.
26 Siegmund, *Was kann und soll die säschsische Siedlungstätigkeit leisten?*, 1.
27 BSV, *Bestimmungen zur Siedlungshilfe des B.S.V.*, 1.

question, but awarded a consolation prize either way.[28] Teutsch was a long-standing supporter of Siegmund's, and the church Landeskonsistorium decided to recommend to its priests and teachers to support the BSV's aims in 1909, and to facilitate its survey of land ownership by disseminating the questionnaires in 1910.[29] Swelling its ranks, various local branches of the Order of Good Templers joined the BSV in 1911, as did the Transylvanian Saxon Farmers' Association (Siebenbürgisch-Sächsischer Landwirtschafts-verein) in 1913.[30]

Although the BSV did not to survive the First World War, Siegmund's agenda is aptly illustrated by the two editions (1914 and 1922) of the flagship volume *The Saxon Population Increase and Defense Book* (a rather rough translation for the German *Sächsische Wehr- und Mehrbuch*).[31] All the more so as the substantial changes in the book's content and language (Siegmund wrote most of the second edition himself) clearly reflect the radicalization of his discourse after the First World War. It is in the 1922 edition that we find a chapter on "The Tasks of Saxon Racial Hygiene," offering a valuable summary of Siegmund's conception of race, deep-seated disdain for mixed marriages, and the designs for a Saxon eugenic fortress:

> The race is the rootstalk from which the trunk, branches, leaves, blossoms, and fruits of national culture sprout. And just as there are different types of rootstalk, so too there are different races and cultures: noble and base ones. If the race is diseased, we can see that in its fruit. If the rootstock dies, the whole plant's life expires with all its blossoms and fruit. We now come to know them, the different forms of racial decay: racial mixing, changing, and degeneration. But how do we prevent it? How are we working to offset it?
> We counter the danger of racial mixing by avoiding racially mixed marriages. No member of our nation should enter into marriage with a non-Saxon. We can make achieving this moral imperative easier for ourselves and our national comrades through the tools of biological segregation [*lebschaftlichen Absonderung*].[32]

28 NA Sibiu, Brukental Inv. 107, LL. 1-29.78, pp. 43–44.

29 NA Sibiu, Brukental Inv. 107, LL. 1-29.81, pp. 1–2. For the questionnaires, see NA Sibiu, Brukental Inv. 107, LL. 1-29.88, and NA Sibiu, CGR, D.XIV.810-19.

30 NA Sibiu, Brukental Inv. 107, LL. 1-29.81, pp. 11, 15–19.

31 Siegmund, Englisch, and Schuster, eds., *Sächsisches Wehr und Mehrbuch*; and Siegmund, ed., *Sächsisches Wehr und Mehrbuch.*

32 Siegmund, "Die Aufgaben der sächsischen Rassenhygiene," 214. For a longer translated extract from this

He thus accentuated national particularities such as language, faith, and customs, while promoting a managed socioeconomic integration (or, better, isolation) of fellow Saxons to uphold cultural barriers and minimize points of contact, with Romanians in particular.

Siegmund also stood at the forefront of the Saxon abstinence movement, and constantly, relentlessly, if hardly ever successfully, he battled Saxon drinking habits in word and print. Having founded the Alcohol Abstinence Society (Alkohol-Enthaltsamkeitsverein) in 1902, Siegmund joined the Order of Good Templars in 1905, and immediately set about creating a Transylvanian branch that eventually reemerged from the First World War, in 1924, as the Romanian Grand Lodge 2 (German), with Siegmund as its Grand Templar.

He was also instrumental to the introduction of school nurses in Protestant schools in 1908, in recognition of which the Saxon bishop Friedrich Teutsch made Siegmund the first "medical member" of the church's governing council, the Landeskonsistorium, in 1920. Siegmund had an ambiguous relationship with the Saxon Protestant church—he complained it was too dogmatic and rigid an institution on the one hand, but courted it for its societal influence and structural network on the other. What resulted was perhaps one of the most remarkable and certainly profitable collaborations of interwar Saxon eugenics, namely the Welfare Committee (Fürsorgeausschuss, FA) which Siegmund created within the church in 1922, and directed unto his death in 1937.[33] The Welfare Committee is profoundly important in facilitating Saxon eugenics, but also speaks to the level of support Siegmund's eugenic agenda enjoyed within the church. It invariably also forced at times a difficult conversation on the compatibility of Christian morality with Siegmund's brand of eugenics (as a moral imperative and a science alike), especially when it came to notions of hereditary talent and the nature of welfare. So, for example, Siegmund's glowing review of Lenz's 1931 *Human Selection and Racial Hygiene (Menschliche Auslese und Rassenhygiene)*, serving to underline the importance of the internal colonization work undertook by the church, while also arguing that "all true science is to be seen as a revelation

text, see Georgescu, "Transylvanian Saxon Eugenics."

33 For a short overview, see Killyen, "Zur Geschichte des siebenbürgisch-sächisischen Fürsorgewessens." For a contemporary German perspective, see Collmer, *Fürsorge als völkische Selbstbehauptung.*

by God, and to work in it is faith-work [*der Dienst an ihr ist Glaubensdienst*]."[34]

A succinct example of what Siegmund considered to be his committee's core purpose is his 1922 "Fact Sheet on the Goals of the Protestant Welfare." In it, Siegmund proposes the committee as a space for collaboration between medicine and faith, all the while referring to the church's secular dimension as a national church, before elaborating on its four main areas of work: "1. Welfare [*Wohlfahrtspflege*], 2. Healthcare [*Gesundheitspflege*], 3. Racial health, 4. Care of the national body [*Volkskörperpflege*]."[35] Quickly dismissing the first two as largely palliative, Siegmund eagerly emphasizes the importance of preventative medicine and of a holistic racial hygiene geared towards safeguarding regenerative, rather than dressing degenerative trends in national health and heredity. That is, the "Protestant welfare wants to create healthy blood amongst our faith brothers, it wants, by working on the ascending, to shorten the descending branch of life as far as possible—yes, even to remove it entirely from the world as far as possible."[36]

Not only did this organization offer Siegmund a novel means by which to pursue his assessments of the Saxon nation's racial health through annual censuses and reports, to seek the implementation of eugenic policies such as the introduction of alcohol- and tobacco-free Sundays, and to coordinate race-anthropological research projects, but it also allowed him to persistently promote his views amongst the clergy, teachers, and wider population alike through the committee's organ, the *Protestant Welfare Worker* (*Evangelischer Fürsorger*), which he edited, as well as through the organization of annual welfare trips and courses, exhibitions, open lectures, etc.

The annual parish reports sent to Siegmund compose a fascinating collage of local realities that afford an insight into how local concerns chimed with the wider eugenic project. Three examples submitted for 1928 illustrate how core themes such as the demographic decline, mixed marriages, and the perceived Romanian competition resonated profoundly. A report by Kürd lamented that while the Saxon community owned sufficient houses, many of these were empty, and that despite this and a declining birthrate, emigration continued to pose a serious problem. Similarly pessimistic, a wel-

34 Siegmund, "Rassengesundheitspflege," 6–7.

35 Siegmund, *Merkblatt über die Ziele der evangelischen Fürsorge*, 1.

36 Ibid., 9. Amongst the plethora of publications extolling the importance of race-hygienic programs to a successful welfare policy, see also Siegmund, "Die wichtigste Aufgabe der landeskirchlichen Fürsorge."

fare report from Klausenburg focused on mixed marriages, and included the author's own graph illustrating their growing prevalence—namely, that 78 percent of marriages in the previous forty years had been inter-confessional, and 48 percent interethnic. A report for Tartlau, for its part, emphasized the perceived degenerative impact wrought by a growing Romanian population: "All these many foreign servants, farm and factory hands here, create an evil proletariat that has a vulgarizing influence on our Saxon youth. [...] Fiscal mismanagement and corruption take hold and threaten to corrupt our community too."[37]

The Welfare Office also composed its own annual reports, and the effect to which Siegmund used these is seen, for example, in his 1926 assessment of Birthälm. Diagnosing Birthälm as "seriously ill," Siegmund set out to explain that it was on the brink of a partially self-incurred extinction due to a combination of a dramatic decline in "racial worth"; the unabated influx of Romanians who were taking over the agricultural sphere while the Saxons had increasingly turned to urban professions; and the degenerative indulgence of alcohol. This was a crisis for which the Saxons of Birthälm were ill-prepared:

> 1. due to a lacking understanding of the nature of disease symptoms; 2. due to their racial incapability [*Untüchtigkeit*], which can be explained, in particular, through a) the consequences of the widespread appearance of swollen thyroids accompanied by idiocy and cretinism (iodine deficiency), and b) the creeping alcohol poisoning [*Weingeistvergiftung*]; 3. as a result of the other known influences of the alcohol plague in damaging health, economy, and morality; and 4. [because of] an insufficient consolidation [*Zusammenschluss*] of the fellow believers as a result of weak leadership.[38]

Siegmund's line of reasoning exemplifies the manner in which he biologized society as a singular patient, a singular body subject to medical scrutiny and treatment—in this case prescribing "re-Saxonification" through internal colonization as the only possible means of saving Birthälm.

The following year, in 1929, there were also three key welfare events:

37 NA Sibiu, CGR. D.XIV.83, pp. 35–38, 39–45, and 55–60, respectively.

38 ZAEV, LK 103.87 (1926): 937, p. 1.

	Sunday	Monday	Tuesday	Wednesday	Thursday	Friday	Saturday
8–9	-	Remit of Protestant welfare	*Lebensraum*	Population/ migration streams	Sick communities	Sick communities	National character and alcohol
$9^{1/4}$–10	-	Noesen	Laws of inheritance	Racial hygiene	Limitating fertility	School hygiene	Youth - Welfare work
$10^{1/4}$–11	-	TB	Are we sinking?	Medical perspectives on alcohol	Furhter education movement, Co-operative banking	Displacement in the life of the Saxon nation	Current goals of Protestant welfare
$11^{1/4}$–12	-	The welfare week	Marriage counselling	Self-Help	Fertility limitation	National (*völkisch*) education	
$12^{1/2}$–$13^{1/2}$	-	Communal lunch					-
4–5	-	Venereal diseases	Cancer prevention	Tour of the Welfare Exhibit	-	Tour of the Welfare Exhibit	-
5–6	-	-	-		Goiter prevention		-
7–8	Communal dinner						
$8^{1/2}$	Welcome			Discussion	Sick communities	Discussion	

Table 1. Program, Bistritz Welfare-Training Course (July 7–13, 1929). Source: ZAEV, LK 103.90 (1929), 19: p.5.

the welfare trip (*Fürsorgefahrt*) to the Bistritz parish (July 10–17), the welfare training course (*Fürsorgelehrgang*) in Bistritz (July 7–13), and the welfare week (*Fürsorgewoche*) (November 3–10). The welfare week on the topic of "creating a healthy life" aimed to raise awareness about tuberculosis and raise funds for a sanatorium, and was certainly a concerted effort, featuring exhibits, a total of 230 medical public lectures across the country, from Bucharest to Czernowitz to Temeschburg, and even a specially commissioned play, *Too Late*,[39] that enjoyed some twenty-nine performances in various parishes (in some cases, such as Mediasch, up to ten shows).[40] However, it is the weeklong training course that is of particular interest here; its

39 Hermann-Mueller, *Ze Spet.*
40 Siegmund, "Bericht über die im Zusammenhange," 75–76.

program was dominated by the usual themes, ranging from displacement to racial hygiene, as well as the Self-Help.

Amongst the twelve speakers at the course there were several key figures, such as Csallner (lecturing on heredity, racial hygiene, child limitation, Self-Help), Misch Bergleiter (further education movement/Volkshochschul-Bewegung and cooperative banking/Raiffeisenhaus), Michael Englisch (education and "Are we sinking?"), Father Dr. G. W. Seraphin (marriage counselling), and the future *Gauleiter* of Siebenbürgen, Dr. Julius Waedt (tuberculosis).[41]

The late 1920s crucially marked a period of substantial, if ultimately troubled collaboration between Siegmund and Fabritius. Having become a Self-Help member in 1927 after a public lecture by Dwovsky in Mediasch,[42] the Welfare Committee also joined in 1928 (but eventually cancelled its membership in 1935), and paid 200,000 lei into its Self-Help account to access a total of 700,000 lei in loans for Moritzdorf, Niederneudorf, and Kuschma.[43]

Siegmund's long-standing flagship project, though, was the Welfare Committee's National Health Exhibit (Fürsorgeausstellung Volksgesundheit), which he modelled on the 1911 Dresden Hygiene Exhibit.[44] Siegmund received final permission from the Landeskonsistorium to begin curating the exhibit in 1934, and tasked Richard Frank, who was on a study trip in Germany in 1934, with negotiations with the Reich Committee for National Health Services (Reichsausschuss für Volksgesundheitsdienst) and Hygiene Museum that subsequently donated ten boxes of exhibits.[45]

The exhibit was ceremoniously opened in the presence of Bishop Glondys and his wife, who was also the exhibit's patron, with Siegmund's opening address insisting that visitors must take away an understanding of welfare as a science, and that "actually only work on ascending life is true, constructive, growing work [*Mehrungsarbeit*]! Today's Germany also makes this sharp distinction, trying to inhibit the continued existence of racially

41 ZAEV, LK 103.90 (1929), 19, p. 4.

42 "Nachricht," 4.

43 Interestingly, the committee itself only contributed a fraction of this amount, with the Women's Society contributing 150,000 lei and Csallner's Society of Child Enthusiasts 32,000 lei. See ZAEV, LK 103.96 (1935), 3217.

44 Siegmund was utterly enthused by the exhibit and wrote a glowing report on his visit. See Siegmund, "Bericht des Stadtphysikus."

45 ZAEV, LK 103.95 (1934), 772; and ZAEV, LK 103.96 (1935), 134.

inferior portions of the nation through sterilization and marriage bans on the one hand and, on the other, striving to raise its race through internal colonization, inheritance rights [*Erbhofrechte*], and support for marriages of the racially valuable."[46]

The exhibit contained some 2,500 items relating to a wide variety of themes, from infant care to school nurses, venereal diseases, race and heredity, physical education, alcohol, tobacco, and cancer.[47] It was undoubtedly a fitting showpiece for a career so thoroughly dedicated to the dissemination of eugenics, one that Siegmund had long fought for, but which he was to relish all too briefly, deceasing in July that year.

Amongst the many obituaries lauding Siegmund's tenacity and achievements, Wilhelm Hager's offers one of the few examples to explicitly connect him to National Socialism: "Siegmund became a pillar, patron, and defender of all constructive [*aufbauenden*] efforts at renewal amongst us. The German nation's departure into the world under the sign of the swastika was, to him, the realization of an old yearning. It is also thanks to his work that this new era did not meet us unprepared. His life's achievement is a bridge from the battles of the ancestors to those of the current time, and an inalienable part of the history of our nation."[48]

A rather more personal account is offered by Richard Bell, incidentally the son of A. Bell, who worked closely with Siegmund in the BSV before they fell out, and who recounts various stories about his childhood experiences with Siegmund. He also aptly contextualizes Siegmund's challenges: "In the beginning, and also for a long time afterwards, people derided him. Called him misfit, fanatic, [and] a stubborn idealist, one who couldn't appraise reality, who saw ghosts in broad daylight. And today? Today we extol ourselves for having had him. Proudly point out that so much of that, what the national socialist worldview brought with it, was already anchored in his wants and works."[49]

What also distinguishes Bell's obituary is his reference to Siegmund's

46 Siegmund, "Ansprache zur Eröffnung," 18.

47 See the three-part article: Siegmund, "Füherer durch die landeskirchliche Fürsorgeausstellung 'Volksgesundheit.'"

48 Hager, "[Obit] Zum Tode Dr. Heinrich Siegmunds," 310. Also see Gyurgyevich, "[Obit] Dr. Heinrich Siegmund"; and Rosenauer, "In Memoriam Dr. Heinrich Siegmund." It is also worth noting here that Siegmund sent copies of his *Wehr- und Mehrbuch* and *Deutschen-Dämmerung* to Hitler in early 1933. See NA Sibiu, Brukental Inv. 107, LL. 1-29.65, p. 76.

49 Bell, "An ihren Früchten," 30.

pragmatic relationship with the church, and while Siegmund was anything but fervent in his beliefs, he certainly found a way of substantially advancing his agenda through it. According to Hansgeorg von Killyen, the most prolific of Saxon medical historians, "today, he is considered to be the founder of modern preventative and social medicine in Saxon Transylvania."[50] This is an equally substantial and consequential legacy, which can shed much light on the various dimensions of not only Siegmund's own eugenic core and the international network that he built around it, but on the wider welfare policies surrounding school doctors, midwives, and public health provisions for tuberculosis, amongst many more.

II. Saxon Racial Anthropology between Berlin and Vienna

A particularly illustrative example of the degree to which Saxon eugenicists collaborated with and benefited from their German counterparts is afforded by the Saxon race-anthropological studies of the 1930s. The manner and extent to which ethnic minorities engaged with the different race-anthropological discourses that courted and studied them is a particularly rewarding field for future research, and the Transylvanian Saxons offer a striking example of these processes in a remarkable episode of professional competition that erupted between Viktor Lebzelter and Eugen Fischer in the early 1930s, when both undertook racial surveys of Saxons.

Virtually all who came to shape and define Saxon racial anthropology and its race-hygienic programs nurtured close ties to the more radical branches of German racial hygiene. Crucially, these affiliations were largely the result of the substantial number of Saxon eugenicists to have studied in Germany, in particular with Eugen Fischer, such as Albert Hermann, Johann Bredt, and Eckhard Hügel. But the realization of extensive race-anthropological surveys ultimately also relied on the Saxon Protestant church, which steadfastly supported the racial research projects pursued by Saxon or external anthropologists through its network of priests and teachers. Beyond the church, probably the most important institutional base of Saxon racial anthropology was Alfred Csallner's National Office for Statistics and Genealogy (Landesamt für Statistik und Sippenwesen) and its Sub-

50 Killyen, "Die Rolle der siebenbürgisch-sächsischen Vereine," 115.

department for Racial Research (Unterabteilung für Rassenforschung). Led by the biology teacher Albert Hermann,[51] the subdepartment was tasked with advancing the biometric study of all Saxon communities (and, interestingly enough, local Romanian and Szekler populations were also included in these studies), with the longterm goal of "striving towards—and directing—the race-anthropological study of the other Germans in this country," as well.[52] It should also be noted that Hermann's assistant Michael Fleischer apparently carried out race-anthropological studies of Saxons in the northeastern regions of Transylvania with a research grant negotiated by Csallner, although further details are currently unknown.

Hermann was one of the few indigenous racial anthropologists to have actively pursued the biometric study of Saxon settlements, and did so with Eugen Fischer's active support and encouragement. Regrettably, Hermann published very little of his anthropometric studies of the Saxon peasantry in southeast Transylvania apart from his landmark 1937 book on *The German Peasants of the Burzenland*. Written for a wider audience, it included a general overview of the region's history, but was grounded in the fieldwork he conducted with the help of local teachers, priests, and peasants in 1932. Having fully surveyed all German inhabitants over sixteen from the villages Heldsdorf, Honigberg, and Weidenbach, Hermann brought the total number of individuals evaluated up to 6,000 by including roughly 20–25 percent of the rest of the region's German settlements as well. Csallner, for one, clearly admired Hermann's work, which he delighted to point out marked the "first larger race-anthropological study to be conducted by an *auslandsdeutscher* researcher, and is likely to be the most substantial race-anthropological study of a homogenous national community to date."[53]

There were, in fact, several biometric surveys of Saxon settlements underway in the early 1930s, even if Albert Hermann's was by far the most comprehensive. Hermann took particular offence with the efforts of the Austrian anthropologist Viktor Lebzelter, demanding his benefactor Eugen Fischer to prevent what he considered to be Lebzelter's attempt to preempt his findings. At times heated, the dispute between Hermann and Lebzelter is partially chronicled by a flurry of letter exchanges dating from 1932 to

51 For a brief biography, see Hienz, "Hermann, Dr. phil Albert."
52 Csallner, "Das Landesamt," 7–8.
53 Csallner, *Die volksbiologische Forschung*, 74.

1935, between Lebzelter, Fischer, and Fischer's protégées, Albert Hermann and Johann Bredt. The church, in its capacity as the research projects' facilitator, is certainly worth taking a closer look at, all the more so considering the wider context of Saxon racial anthropology's heavy reliance upon Fischer's Kaiser-Wilhelm-Institut (KWI) for funding and training. An in-depth investigation of these institutional (and certainly ideological) links is currently sorely missed.

The letters reflect Hermann's unrelenting ambition (as well as his dislike of Lebzelter, whom he considered to be careless methodologically, and whom he reproached for working with too small a sample empirically), and Lebzelter's growing irritation with a newcomer whose "fear I would steal the raisins off his cake"[54] had led him to agree, as a favor to Fisher, to delay publishing his own findings.

When the church issued two directives in support of Lebzelter's planned anthropological research on June 17 and 24, 1932,[55] the response was swift. On July 2, the Bistritz Bezirkskonsistorium submitted a letter to the church protesting that Lebzelter's work would imperil the far more thorough studies funded by Fischer and conducted by Hermann in Burzenland, and Bredt in Noesnergau.[56] Fischer, having been alerted to Lebzelter's plans by Hermann, had already written to Lebzelter on June 28, requesting a clarification of his intentions, and establishing his priority, before asking Lebzelter to change his plans and study a different German ethnic group, such as the Banat Schwabs or Bessarabian German communities: "You do not know that I myself have been to Siebenbürgen with colleague von Verschuer three years ago, have trained a number of teachers in anthropological techniques, and got a national survey [*Landesaufnahme*] underway. Mr. Prof. Dr. Hermann in Kronstadt who, with Father Bredt, then came to my institute for some time to get training, now has the matter in hand and is working towards a comprehensive survey spanning both anthropological measure-

54 NA Sibiu, CGR.D.XIV.836, p. 18.

55 The first was a letter of recommendation encouraging all parishes to aid Lebzelter's research, in particular through the mobilization of volunteers, and especially entire extended families. The second encompassed Lebzelter's *Merkblatt* outlining his aims and methodology, namely to compare Saxon racial profile—tabulated on the basis of height, hair, skin and eye color, the shape of the head, face, nose, and mouth—to that of areas in Germany and Luxembourg which are likely points of origin. See ZAEV, LK 103.94 (1932), 2256.

56 Ibid.

ments and genealogical research in church archives."[57]

While Lebzelter's reply to Fischer has not been preserved, the latter's response to it on July 11, 1932, is certainly more conciliatory, with Fischer seemingly satisfied that Lebzelter's study would offer depth as well as breadth, and hence would not be just the cursory sampling that Hermann and others lamented. Staking his claim more firmly, Fischer also asserts: "Father Bredt has only retired so he could work entirely for us. We pay him the difference between salary and pension. I went to Bishop Teutsch personally. So I do not want to disturb the work you have begun, under the conditions that, first, we protect the study through a clear separation [of areas], and second, we don't compete with each other when publishing [the findings]."[58]

Over the following months, Lebzelter visited Romania numerous times, supplementing his anthropological research on the Saxons whenever his engagements for the Romanian military permitted,[59] albeit on a tight budget given the references to his attempts to supplement it with fees for giving lectures.[60] Lebzelter was surely not the only international luminary to do so, and it certainly needs to be flagged that there is substantial mileage in the study of the plethora of lecture series in the 1930s and early 1940s featuring visiting speakers. Within this context it is not surprising to find that Fischer himself also presented a series of talks on "Race as a History-Determining Factor" in the fall of 1941: in Kronstadt on October 30, hosted by the Research Institute of the German National Community in Romania; in Hermannstadt on October 31; and—surely amongst other as yet unknown events—in Temeschwar on November 3, where Fischer led a session of questions and answers with local teachers organized by the regional Schools Office, before lecturing at the Technical College as the guest of its dean, Prof. Dr. Racoviță (who, for his part, was keen to establish a Romanian anthropological institute modelled on Fischer's in Berlin-Dahlem).[61]

Fischer's proviso about publishing findings reignited the issue the following year, and in a letter to Siegmund reminiscing about the time they spent together, Lebzelter enquired as to whether Hermann had published

57 Ibid.

58 Ibid.

59 See, for instance, Lebzelter's letters to the church dated July 25, October 8 (which incidentally also mentions that he spent two enjoyable evenings with Bredt), and October 26, 1932. Ibid.

60 See, in particular, Lebzelter's letters to the church dated October 8 and 24, 1932. Ibid.

61 See Lammert, "Rasse als geschichtsbildender Faktor."

anything yet: "After my return to Vienna last autumn, Prof. Fischer - Berlin [. . .] [requested] that I don't publish any of my Saxon material until Johann Bredt and Prof. Hermann published something. The latter had threatened he would cease his work if something of mine appears. Now that is really too much posturing [*etwas arg possierlich*]."[62] Insisting there was room for all of their research projects, and that he had no interest in stealing anyone's thunder, Lebzelter asked Siegmund to exert his influence on Hermann, whom he did not know personally.[63] Siegmund then duly wrote to Hermann, whose reply started with a rigorous defense of why he asked Fischer to "protect" his work, and lamented that while Lebzelter had acquiesced to Fischer's request to not research in the Burzenland, Lebzelter had failed to contact him while in Kronstadt in the fall of 1932. Interestingly, Hermann also mentioned that Eckhard Hügel had recently returned from his studies in Berlin with Fischer, and noted his plans to study the areas around Agnetheln and Schässburg: "I am of the opinion that these home-grown ambitions should be supported, and I will ask Prof. Fischer to support them."[64] Siegmund, for his part, and having in turn consulted with Bredt, informed Lebzelter that Hermann hoped to publish in 1934, and that he himself could not intervene, as the church had awarded Hermann privileges that he—as a member of the church governing council—needed to respect, despite wanting to see Lebzelter's work published.[65] Lebzelter was obviously not enthused by this response, insisting that he had no interest in a "polemical" confrontation with Hermann, and that Fischer could not actually prevent him from publishing.[66] By 1935, with Hermann still several years away from publishing his findings, the dynamic seems to have changed when we find Lebzelter writing to Siegmund to ask whether Saxon publishers would be interested in his research, as Fischer, to whom he had recently spoken with in Berlin, had seemingly lost interest in Hermann and hence now felt free to go to print.[67] It is unclear what happened in these two years, but both Lebzelter and Hermann eventually published their findings in 1936, albeit Hermann's contribution (which, it is interesting to note, was published in *Klingsor*) was an

62 NA Sibiu, Brukental Inv. 107, LL. 1-29.65, p. 38.
63 NA Sibiu, CGR.D.XIV.836, pp. 18–19.
64 Ibid., 20.
65 Ibid., 21.
66 NA Sibiu, Brukental Inv. 107, LL. 1-29.65, p. 37.
67 Ibid., 39.

abstract of the book released the following year.[68]

Johann Bredt similarly carried out race-anthropological research under Fischer's guidance. Bredt had studied theology and history, along with natural history and chemistry, in Tübingen, Klausenburg, and Leipzig between 1892 and 1896. In many ways similar to the career path Csallner had taken, Bredt spent the following seven years teaching in Hermannstadt and Bistritz, before switching to the priestly profession. He was subsequently employed by the villages Oberneudorf (1903–1906), Waltersdorf (1906–1912), Windau (1912–1919), and Kleinbistritz (1919–1930), before "retiring" and resettling to Bistritz in 1930 to pursue the anthropological research he was ultimately unable to complete because of an unnamed illness Bredt battled, but lost to in March 1936.

Bredt's core contribution to the Saxon eugenic discourse were his studies of "the national body" (*Volkskörperforschung*). As Johann Bredt's son Heinrich put in an obituary to his deceased father, Bredt's methodology was in many ways a logical progression from Csallner's sacralized statistics, in that it sought to reveal the "real causes of the Saxon nation's looming biological death" by replacing generic "statistical overviews with a living picture of tribes and families."[69] Bredt aspired to create a "transparent" nation through not merely the study of family trees, but the anthropo-genealogical excavation of entire towns. In other words, he studied the historical evolution and spatial variations of particular settlements in order to produce precise surveys of any given settlement's economic and demographic status at any given point in time (essentially, like the cross sections of tree rings).

Csallner was not the only one to see the biopolitical utility of a statistical system offering such "scientifically tailored" evaluations of any respective town's struggle against displacement and degeneration as evidenced over time, and the Stiftung für deutsche Volks- und Kulturbodenforschung in Leipzig actively encouraged Bredt to publish a summary of his conceptual approach and methodology in 1930.[70] This fifty-five-page-long pamphlet, along with his landmark 1929 case study on *Windau in Past and Present*[71] constitute the bulk of his short but intense career, one that ben-

68 Lebzelter, Wastl, and Sittenberger, *Ein Beitrag*; Hermann, "Die Deutschen Bauern des Burzenlandes."

69 Bredt, "Volkskörperforschung," 141.

70 Bredt, *Volkskörperforschung.*

71 Bredt, *Windau in Gegenwart und Vergangenheit.* A revised version was later published by Günter Litschel: Bredt and Litschel, *Windauer Heimatbuch.* Bredt also published a handful of shorter pieces, the two most im-

efited immeasurably from probably the most prolific donor and patron of Saxon eugenics, the director of the Kaiser Wilhelm Institute for Anthropology, Human Heredity, and Eugenics, Dr. Eugen Fischer. In fact, Bredt's "retirement" from priesthood in 1930 was indebted to a five-year research grant from the Notgemeinschaft der deutschen Wissenschaft that carried an annual stipend of 3,000 RM, which had been secured by Fischer. In a letter sent to the church LK in January 1930, Bredt elaborated on his plans to comprehensively study the Nösner region (including Bistritz), a vast project he would only be able to accept if freed from his priestly duties, and under the condition that the church—having recognized the seminal importance of his work—continued to pay him a salary.[72] Bredt subsequently began his research and, remarkably, even augmented his genealogical data with a biometric study of the rural areas surrounding Bistritz, which regrettably he was unable to evaluate and publish prior to his death.

Although Bredt's work was, to a lesser degree, continued by his student Michael Fleischer, Csallner was left likening Bredt's *Volkskörperforschung* to an abandoned orphan when he lamented: "And so Bredt had something to offer not merely to our Transylvanian Saxon science, but to the great German science. It is not his fault that he was unable to complete it. But where our Transylvanian Saxon science does not have the strength—at least not for the time being—to finish it, maybe the great German science will complete and pick up the gift!"[73]

In addition, there was also a range of small-scale race-anthropological studies conducted by local priests and teachers. A brief, indicative example of this is offered by Arz von Straussenberg, who having read Hans Guenther's books felt inspired to study the prevalence of "Nordic markers" in Hermannstadt, and subsequently assessed ca. 300 people (157 teachers and 137 young farmers) in 106 settlements as to their head and facial indices, eye and hair color, height, weight, and chest circumference. Many of the conclusions he reached correlated directly with core eugenic themes, such as the insistence upon the heightened hereditary worth of the teaching and priestly professions, when Straussenberg asserts that 85 percent of the teach-

portant of which are Bredt, "Der Schwindsucht in unserer Gemeinde"; and *Vererbungskunde und Katechismusbehandlung.*

72 NA Sibiu, CGR, D.XIV.831, pp. 69–70.

73 Csallner, *Die volksbiologische Forschung*, 190–91.

ers and priests had a facial index of ninty or above, compared to the overall average of 65 percent: "They hence really constitute a spiritual and Nordic elite [*Auslese*], which can be seen in the bulk of their sons."[74] Unsurprisingly, Straussenberg continued by deploring how these valuable and talented groups were facing extinction due to child limitation, imploring: "How much Nordic blood have we already internally lost in this way? How much more Nordic blood must be lost until this enters the national consciousness, until it comes to its senses and proceeds to build a new community?"[75]

III. The "Child Enthusiast" Alfred Csallner

Csallner was clearly a central figure in Saxon interwar eugenics, coming to the fore during the interwar period by not only assuming a pivotal role in verbalizing and spreading eugenic ideas, but also because of the shift from clerical and political means of translating these into practice that accompanied the Self-Help's rise to power.

Figure 2. Alfred Csallner (ca. 1940).

74 Straussenberg, "Bericht über Körpermessungen," 46.
75 Ibid., 47.

Csallner's eugenic ideology was ultimately defined by Siegmund's own growing body of work, as well as by the close friendships he maintained with Wilhelm Schunn and Johann Bredt. Csallner argued that the Saxons had unexpectedly found themselves in the advanced stages of a cataclysmic and apocalyptical degenerative crisis that had simmered unnoticed for decades prior to being recognized and diagnosed by his mentor Siegmund: "This danger of degeneration is the greatest of all threats. We, of course, don't see it yet, not even our leaders want to recognize and acknowledge it. And we who have must therefore be all the braver and call all the louder."[76] And call to arms he did. Csallner, as brief as his career might have been, unleashed a torrent of publications, set up and ran several organizations, and he even resorted to penning poetry and plays in a bid to enlighten the Saxon body politic to the existential threat they faced. However, Csallner—at heart always a priest—also fervently believed in the clergy's duty to lead and shape their parishes, and he dedicated a significant portion of his time to lobbying the church to embrace the race-hygienic measures he deemed alone capable of turning this impending national disintegration into regeneration, a national rebirth.

Csallner's eugenic ideology underwent an ongoing process of politicization and radicalization throughout the 1920s and 1930s. As evidenced by the various eugenic proposals he tried to translate into reality through the church, which he and Siegmund had so desperately tried to reform from within, Csallner was to demand increasingly centralized and authoritarian measures be taken to contain and ultimately expel the single- and two-child family from Saxon national life, to expel those spiritually bankrupt elements that could (not to mention would) betray their heritage by entering mixed marriages, and to expunge the invading ethnic others who had inflicted such heavy damage upon the Saxon economic sphere and the integrity of the Saxon *Lebensraum*.

In 1927, Csallner founded the Society of Child Enthusiasts (Verein der Kinderfrohen, or VdK) that earned him his nickname. With stringent membership criteria demanding that applicants already have four children (or promise to do so in the near future), it is little surprising that the VdK failed to attract the legions Csallner had called to his movement's banner. Consequently, frustrated by his inability to implement eugenic policies through

76 Csallner, "Die grösste Gefahr," 257.

either the church or his Society of Child Enthusiasts, Csallner turned to the ideologically far more sympathetic Self-Help in 1929. He inadvertently became a central figure in a rapidly growing movement that understood racial hygiene as a "divine gift" towards regenerating an idealized Saxon nation. Csallner exerted a considerable influence over the design and implementation of Schunn's neighborhood population policies, while simultaneously pursuing his own vision of a eugenic fortress—an impressive engineering project that humbly began as the Self-Help's Race Office in 1932.[77]

Although it was rather a more virtual than tangible institution, the Self-Help's rise to political power in 1933 allowed Csallner to transform his Race Office into the considerably more influential National Department for Statistics, Population Policy, and Genealogy (Landesarbeitsstelle für Statistik, Bevölkerungspolitik und Sippenwesen, or LSBS). When the Verein für das Deutschtum im Ausland (VDA) encouraged him to lay down his priestly robes and offered to pay his salary in 1936,[78] Csallner was delighted with the prospect of finally dedicating himself to his work full time. 1936 was a tumultuous year in Saxon politics, following the descent of the Self-Help movement into civil war in 1935, a reality that was anything but favorable to conducting the sweeping statistical studies (that both by their nature and design required large numbers of helpers) so characteristic of Csallner's work: "In the first couple of years, tied down by my main profession, and despite my best efforts to extend my work to include all districts [*Gaue*], I could almost exclusively work on and for Transylvania—until in 1936 I was finally able to resign from my pastorate and dedicate myself entirely to my *völkisch* work. I was, naturally, then inhibited by the internal divisions: many of those with whom I would have wanted to work stood on the other side of the fence and not only refused to collaborate, but in some instances attacked my work."[79]

77 See Csallner, "Das Landesamt," 1: "The National Office for Statistics and Genealogy has grown out of the *völkisch* work I began towards the end of 1918, and have pursued out of my own conviction in diverse directions, and with an expanding scope as a private man. In 1932, I approached the leader of the Self-Help, our current national leader Fritz Fabritius, with the suggestion to create a Race Office of the Self-Help, and offered to direct it as a honorary position myself. Fabritius accepted my proposal, and entrusted the management of the Race Office to me."

78 Csallner, *Meine wissenschaftlichen Arbeiten* [1975], 7; and Csallner, *Die volksbiologische Forschung*, 77. In fact, Csallner enjoyed some substantial financial assistance from Germany. He received funding from the Deutsche Forschungsgemeinschaft and the Südostdeutsche Forschungsgemeinschaft, as well as from the Verein für das Deutschtum im Ausland.

79 Csallner, "Das Landesamt," 1.

By 1938, the conflict was resolved through Berlin's intervention, and Csallner was to reach the pinnacle of his political influence and literary output with the reorganization and expansion of his department into the National Office for Statistics and Genealogy (Landesamt für Statistik und Sippenwesen, or LSS). It was Csallner's main means of proliferating, and partially even implementing his eugenic agenda prior to Andreas Schmidt's arrival and the Third Reich's *Gleichschaltung* of Romania's German minorities in 1940. Although the LSS was given the status of an institute briefly thereafter, Csallner soon fell out with Schmidt, who proceeded to rid himself of Csallner by briefly sending him to Germany before dismantling the LSS in 1943. A year later, in 1944, Csallner found himself arrested and interred in the Romanian Târgu Jiu and Turnu Măgurele prisons.[80] Upon his release in 1946, Csallner returned to his priestly robes and picked up the pen again—this time, though, to write novels and a series of short stories. Following two further bouts in Romanian prisons and a string of interrogations and house searches, Csallner immigrated to Germany in 1974, where he died at the age of 97, in 1992.

Of and on origins

Born to Dr. Ludwig and Frieda Csallner in Bistritz in 1895, Alfred Csallner, like virtually all of his peers, studied and spent the First World War as a student[81] in Budapest, Vienna, and Berlin,[82] before hastily returning to Transylvania to tend to his sick father, in May 1918. Due to this premature return to Bistritz and the dissolution of Austria-Hungary that followed shortly thereafter, Csallner was never to complete his degree.[83] He did, however, publish an abstract of sorts in 1919, on "Landownership in the Nösnerland" ("Die Grundbesitzverhältnisse im Nösnerland"), the first in a long list of publications grappling with the continuous loss of Saxon *Lebensraum*. Also in 1919,

80 Csallner, *Meine wissenschaftlichen Arbeiten* [1975], 8.

81 Csallner reflected on the decision not to enlist in the 1923 version of his CV. He insisted that bowing to his parents' wish to make use of the provision for theology students to be relieved of duty had left "a weight on his heart that never left," and he was equally adamant that with his "youth's eagerness and enthusiasm" unsatisfied, his second-year studies "where the most horrible times that I have ever had. I became wholly ill." See ZAEV, LK 103.84 (1923), 5647, p. 2.

82 Csallner both began and ended his studies in Budapest. Having enrolled there for the full academic year 1913–14, he subsequently spent three semesters in Berlin and a fourth in Vienna, before returning to Budapest for the first semester of 1917–18. Ibid.

83 Csallner, *Meine wissenschaftlichen Arbeiten* [1944], 2.

he published a characteristic piece on declining fertility rates unto 1876, "On the Saxon Nation's Development in the Nösnerland" ("Zur Entwicklung des Sachsentums im Nösnerland"). Perhaps the most interesting facet of this early example of Csallner's work was that, in his own words, "around 1919 I still believed I had to defend Saxon women against too damaging criticisms"—that is, before he "realized" the extent of the crisis which falling birthrates implied.[84]

That, clearly, was not to happen again in a career dedicated primarily to the statistical analysis of the degenerative impact that single- and two-child families had on the nation's spiritual and racial constitution, and upon what he deemed the hereditarily best endowed in particular. His career, in Csallner's own words, owed its path to some degree to the impact of his father's death, shortly followed by the cataclysmic conclusion to the First World War, a time when Csallner was to find a new sense of purpose in the embrace of the *völkisch* work he had begun during his studies: "When the total collapse at the end of 1918 and its aftermath had violently shook and profoundly churned me, I—filled with such hunger for action and work—threw myself into all kinds of work for our nation. It was a new life that had begun for me, a life that I considered really worth living, and which I was happy for."[85]

By way of a brief overview, Csallner worked as a teaching assistant (*Supplent*) in Bistritz for a high school and a girls' school between 1918 and 1922, before moving to teaching in a vocational college (*Lehrerinnenbildungsanstalt*) in Schässburg in 1923.[86] Crucially, Csallner was then ordained a priest and successively employed in Roseln (1923–29), Stolzenburg (1929–32), and Kleinscheuern (1932–36) before discarding his priestly robes in favor of pursuing his eugenic cause full time in 1936. His departure from Kleinscheuern in April 1936 descended into chaos, as had been the case in Stolzenburg,[87] with Csallner left fighting an increasingly bitter, almost year-

84 Csallner, *Die volksbiologische Forschung*, 80.

85 ZAEV, LK 103.84 (1923), 5647, pp. 2–3.

86 Ibid., 3.

87 The reasons for Csallner's short stay in Stolzenburg are recounted by Wilhelm Klein when he writes that "as his salary was not paid for some time, Csallner left the community Stolzenburg, became the priest of Kleinscheuern, and sued Stolzenburg for the payment of his outstanding salary. To ensure he got his money, he pawned the valuable church carpets. Considering the enormous scandal this action evoked, the Landeskonsistorium saw itself forced to intervene, and the auction was stopped." See Klein, "... lies Kirchenteppiche pfänden," 4. For the correspondence regarding the carpets and subsequent disciplinary action against Csallner, see ZAEV, LK 103.97, 618, pp. 1–11.

long struggle to have the village presbyterium pay a growing backlog of outstanding wages.[88] The immediate cause for the conflict was an application for a year's leave from his priestly duties (submitted in July 1935), which he hoped to secure while still planning the Race Office's expansion into the LSBS. Towards justifying his request, Csallner offered an interesting abstract of his institution's purpose when he wrote: "The task which National Socialism has set itself in Germany, namely the biological salvation of the mortally endangered German nation—that will now also be tried amongst the even more threatened Germans in Romania with a special office. I am supposed to lead this population policy office directly attached to the Central Committee of the Association of Germans in Romania, and will therefore have to resign my pastorate and leave the church."[89]

Csallner was obsessed with genealogy, in that it served to visualize how anyone was a mere cog in the grand clockwork of time. Genealogy allowed any given individual (and, by extension, the community at large) to deduct and define his or her particular talents (and, again by extension, also his or her inherited deficiencies), and to thereby determine not only career choices but also the choice of spouse and family size. Genealogy was hence to become the keystone of Csallner's definition and solution to the perceived crisis, and a crucial tool to his bid to instill eugenic thought amongst the masses.

This unwavering insistence on "nature over nurture" and a concise summary of the methodology of his study of hereditary talent are afforded by his article entitled "On the Importance of the Choice of a Spouse to the Nature and Destiny of Children," published in 1934.[90] It appeared as a journal article in *Der Aufbau*, but more interestingly still, it was published as the inaugural issue of Csallner's own pamphlet series, *Nation and Race—Nation and Space* (*Volk und Rasse—Volk und Raum*), the official organ of his Department of Hereditary Biology (Amt für Erbbiologie) housed in Karl von Möller's Cultural Institute of Germans in Romania (Kulturamt der Deutschen in Rumänien in Hermannstadt), founded in 1934.[91]

88 Scattered throughout the file ZAEV, LK 103.97 (1936), 58.

89 Ibid., 9.

90 Csallner, *Von der Bedeutung der Gattenwahl.*

91 Not to be mistaken with its namesake founded by Richard Csaki in 1922 to advance a broader cultural exchange between all of Romania's disparate German communities, which folded due to financial pressures in 1931, and following which Csaki was appointed director of the DAI in Stuttgart in 1933, cementing his role as a key link between Saxon eugenic research and institutions of the Third Reich. On Csaki's Kulturamt, amongst others, see Ciobanu, "Relațiile culturale."

An organization thoroughly suffused with eugenic ideas and programs, the first issue of the Kulturamt's organ, *The Construction* (*Der Aufbau*), featured a remarkable call to race-hygienic arms by Möller: "Our German *Weltanschauung* entirely stems from the racial idea [*Rassegedanken*] and the knowledge of hereditary biological laws. These, hence, stand at the forefront of every *Kulturamtwalter*'s work. He therefore has to become acquainted with racial sciences and hereditary biology, and to possibly also own the key books underpinning them."[92] That said, the first, five-day-long training course for its officials convened in Mediasch in July 1934, and featured a familiar list of speakers, including Albert Hermann (director of the Department for Race, or Rassenamt) lecturing on race, and Csallner on hereditary biology, couched between talks by Möller, his deputy Misch Orend, and Heinrich Scheiner amongst others. The course concluded with a trip to the NEDR's folk festival.[93]

Consequently, the next journal issue featured a new subsection on "Racial Questions, Hereditary Biology, Eugenics, and *Sippenkunde*" ("Rassenfragen, Erbbiologie, Eugenik, Sippenkunde") that came to publish a number of Csallner's articles, and offered one of the few uses of the term "eugenics" in contrast to racial hygiene. Elaborating on this area's rapid expansion, Möller believed that the continuously worsening state of the Germans in Romania amounted to a "holy necessity" to assume a scientific and propagandistic approach to solving a core problem—"namely, the battle against the hereditary-biological and race-hygienically wrong lifestyles of entire tribes [*Stämme*]."[94] With this he meant, in particular, the Banat Schwabs, and Möller publicized Fritz Klingler's research on the Banat's racial and demographic constitution, dedicating a special issue in 1934 to the question "Is the Schwab Dying?" ("Stirbt der Schwabe?").[95]

92 Möller, "Behelfe für rassische Aufklärungsarbeit," 21. Tellingly, Möller's prescribed reading list of seven books consisted of works by Hans Günther, Hermann Siemens, Philaletes Kuhn, Dr. von Leers, Friedrich Scheumann, and Martin Staemmler. For his views on eugenics, see also Möller, "Erbsünden," which decries both couples with hereditary diseases entering marriage and the declining birthrates as "hereditary sins"; and Möller, "Erbbiologischer und eugenischer Wegweiser."

93 For the full program, see Möller, "Arbeitsordnung."

94 Möller, "Der Aufbau."

95 See, in particular, Klingler, "Die Bevölkerungsbewegung." On the impact of mixed marriages in the Banat, see Hoffmann, "Völkische Mischehen im Banat." For a more useful summative rather than critical review, see Harasser, review of "Völkische Mischehe im Banat."

Returning to Csallner's article, in it he compared the final report cards of 176 peasant children in 14 Saxon rural communities to those of their parents, to provide clear evidence for the inheritance of not merely academic skill, but spiritual worth. Working from the assumption that "our heredity remains unchanged," Csallner asserted that report cards convey an impression of any given student's "physical health, ambition, and conscientiousness," in a bid to prove that "the better a father's or mother's school achievements, the better—on average—those of their children, and vice versa."[96] This thesis, in its turn, justified Csallner's demand that genealogy should be taken into consideration when it comes to "choosing" a marital spouse: "If we just open our eyes, and become accustomed to viewing every individual as a limb of the tribe [*Sippe*] from which he has grown, then we can evaluate his overall hereditary worth and predict quite accurately— indeed, frequently definitively—what his children are going to inherit from him, how their self and destiny, their worth for the community is determined by him."[97]

It is this remarkable, repeated insistence on the unequivocal laws of heredity, and the belief that virtually any aspect of one's soul and biological worth is predetermined by genealogy that, equally unequivocally, guided Csallner to the logical conclusion: "The more we know to tell ourselves about these primordial laws of life, whether we are even allowed to conceive new life or not, and with whom creating this new life would be a crime, with whom a joy and blessing—all the more will we be able to avoid disaster, let less of the good rot, all the purer and nobler will the noble remain, and all the better the good become."[98]

Csallner launched a number of different bids to encourage the general public's engagement with racial hygiene, from lecture tours to perpetual public calls for assistants.[99] A particularly illustrative example relating to genealogy as a means of introducing the wider public to the dictates of heredity came in the form of an article published in *Volk und Heimat* in 1938, an open call for laymen to become citizen scientists.[100] Revisiting a

96 Csallner, *Von der Bedeutung der Gattenwahl*, 2.

97 Ibid., 1.

98 Ibid., 2.

99 For an example of Csallner's frequent offer to mentor, and in this case even provide the relevant data to anyone willing to study the "national biological" state of Saxon rural settlements, see Csallner, *Über volksbiologische Forschungen*.

100 Csallner, "Laien in die volksbiologische Forschung!"

training course he had organized in 1928 to introduce to the teachers and priests of the Schenker church parish the various "national biological" studies that they could undertake themselves, this article offered ready-made templates for four studies that focus on the academic professions' reproductive patterns in relation to inherited talent. The first was a study identifying heredity trends in relation to "Families with Long and Short Life Expectancies," while the second emphasized the interdependence of "Hereditary Traits and Honorary Positions" by correlating honorary posts with their holders' school grades. The overall results expected from the latter study are somewhat apparent from the title he gave the third one, namely "School and Lifetime Achievements." Finally, the fourth study problematized "The Emigrated," and sought to establish the extent to which emigration was particularly prevalent amongst the hereditarily best endowed.

It is clearly little surprising that given this approach to the dictate of genealogy, Csallner had published, as early as 1927, a sixteen-page-long pamphlet outlining his own system of designing and labelling a family tree.[101] Interestingly enough, he chose his own family as the example by which to illustrate its methodology, and amongst the 345 ancestors (spread over forty-six generations)[102] that he painstakingly traced, one also finds several of the personalities that were to shape the early stages of his eugenic discourse. One such example is his grandfather Karl Simbriger, whom Csallner idealized for turning the "poor and alcohol-drenched Rottenholz into a sober and wealthy village."[103] Indeed, Csallner repeatedly wrote about his childhood ambition to pursue priesthood precisely because he thought the church offered the best way to "serve" his Saxon nation in crisis: "I believe to remember that when asked what I would want to become, for a long time I had always answered that I would first be a professor, then a priest in Rottenholz, and finally a bishop. And even when, as a teacher, I again felt the urge to become a priest—but wasn't sure whether my ideals would let me—even then it was always Rottenholz that spurred my fantasy. And I did become a priest, but not in Rottenholz."[104]

101 Csallner, *Sippschaftstafeln*, 2.

102 Csallner, *Zur Verleihung des siebenbürgisch-sächsischen Kulturpreises 1982*, 1.

103 Csallner, *Rottenholz*, 82.

104 Ibid., 32–33.

On a final note, it is interesting to find Karl musing about Siegmund's church reform agenda towards the end of his autobiographical *Rottenholz* book. With this conversation set in 1916, Karl's evaluation of Siegmund's ambitions as remembered by Csallner is remarkably illustrative of the Saxon eugenicists' overall quest to recruit the church to their cause. What is more, it offers an early indication of the secular population-policy ends the church was eventually to serve, while alluding to the "torchbearer" role its priests were assigned in lighting the flame of spiritual rejuvenation necessary to implement and engrain it:

> Read what the young Mediasch doctor Dr. Heinrich Siegmund wrote about our church: He'd love to break it up, as he believes that its dogma—even that of our liberal Saxon church!—is stubbornly clinging to evidenced errors and is setting unreasonable demands. And yet he admits that his love for our nation has taught him to love our church. And so, instead of destroying it, he wants, in his own words, to force the seemingly impossible and bridge the contradictions between scientific conviction and deeply pious intolerance. He even wants to increase the church's esteem, influence, and power—to expand it inward and outward, and fashion it into an even stronger national bulwark.[105]

Csallner was certainly to latch onto his grandfather Karl's example of the regenerative influence that a "properly" minded and trained priest could exert upon his flock. At the same time, he also consistently strived to reform the church's structure and mission (be that secular or spiritual) from within, along the same lines Siegmund had trodden.

The ideological divide: Heinrich Siegmund, statistics, and Lebensraum

What ideological influence did Siegmund exert over Csallner, and what lessons did the latter learn from him in the quest to popularize and proliferate his own eugenic views? In short, Siegmund's influence on Csallner was absolutely seminal. Following their first encounter during a train ride in 1922, Siegmund introduced him to the German race-hygienic literature that was to form the backbone of his understanding of ethnic hierarchies,

105 Ibid., 124–25.

hereditary infallibility, and their combined dictate of socioeconomic population policies. The extent to which Csallner felt himself indebted to his mentor—who was so persistently either ignored or ridiculed by his contemporaries—is evident, if nothing else, in the astonishing thirty-seven-page-long homage to Siegmund dominating his 1940 book on *"National-Biological" Research amongst the Transylvanian Saxons.*[106] This, as he called it, "monument" to Siegmund's achievements exhaustively covered Siegmund's eugenic career and the numerous turbulences it persisted through (while, interestingly, making hardly any mention of the Land Preservation Society). What is more, albeit in passing, it touched on their collaborate work in the church's Welfare Committee and in its Settlement Office (*Siedlungsausschuss*), and on their activities in and through the International Order of Good Templars. It also revisited and apologetically justified the conceptual rift that had begun to widen between Siegmund and him in the 1930s.

Tellingly, Csallner credited Siegmund with primarily two core achievements: first, for having been the first to understand the Saxon nation as a living biological organism, and for thus being able to feel its pulse and to diagnose its ailments in biologically definite, verifiable, and absolute terms. Second, Csallner lauded Siegmund for having not merely understood the nature of the Saxon disease, but for his tireless quest to awaken his peers to it, regardless of public opinion. One clearly hears a subtle undertone of admiration and anger in Csallner's account of Siegmund's perpetual humiliation by a status quo increasingly annoyed with this overly "alarmist" and overtly vocal "pessimist":

> He [Siegmund] clearly burdened his countrymen and their national leaders too heavily with the responsibility for the realization of these possibilities—he had asked for too much, for too great a change, to relinquish and self-sacrifice too much. And because this made him uncomfortable, they simply discounted all the rich possibilities Siegmund believed in. Clearly, only pessimism remained. But as this was in and by itself unpleasant, one mostly

106 Csallner, *Die volksbiologische Forschung*. This book was the crown jewel of Csallner's overall body of work, and provided a road map to the eugenic discourse as he understood it to have evolved and matured. It also furnished a particularly valuable grand summary of his own work (if not the gradual politicization and radicalization of his views). Published as the fourth volume in the book series Contributions to the Understanding of Germans in Romania (Beiträge zur Kenntnis des Deutschtums in Rumänien), edited by Rudolf Spek, this book gained an extra significance in that it was addressed to a primarily non-Saxon audience. For a glowing review, see Harasser, review of *Die volksbiologische Forschung*.

> denied that as well—argued against it all the more eagerly as Siegmund persistently wanted to burden one with the responsibility for the looming demise of the Transylvanian Saxons, and became a nuisance.[107]

Csallner understood the Saxon status quo's dismissal of Siegmund as primarily driven by their unwillingness to adapt to the realities of the dysgenic crisis. Siegmund's insistence thereupon had resulted not merely in a widespread apathy towards his work, but in outright hostility.

Csallner clearly admired Siegmund for having produced a medicalized discourse on the Saxon nation that defined its various diseases in biological terms, but to what extent did their visions overlap? Csallner, if in a less intense form, wholeheartedly agreed with Siegmund that the Saxons were spending a disproportionate amount of their national wealth on alcohol and tobacco, money that should have been much rather deployed towards recapturing *Lebensraum*. Csallner entirely subscribed to Siegmund's idealization of the racially pure peasant as the quintessential foot soldier in the quest to defend the Saxons from the ever-growing Romanian "flood." However, they certainly did not agree on the core cause of the Saxon demographic decline, and Csallner eventually saw no other choice than to publicly disagree with and discount Siegmund's theses on the links between a dwindling Saxon fertility rate and its *Lebensraum*, together with the methodology by which he had arrived at them.

According to Csallner, Siegmund had arrived at two erroneous conclusions. He ardently believed that wealthier families were likely to have more children, and poorer families less. This, in turn, implied that Saxon settlements were "overpopulated," an inverse dysgenic crisis that could only be alleviated by a vigorous program of territorial expansion to increase the landed wealth of poorer families, which would subsequently also bear more children. In other words, Siegmund understood the decline in Saxon fertility not as the cause of a perpetual loss of *Lebensraum*, but as its logical, Malthusian conclusion. These views clearly contradicted those advanced by Csallner, who, considering the significant institutional influence Siegmund exerted through his Welfare Committee, deemed them dangerous to his own agenda. As Siegmund repeatedly refused to change his position,

107 Csallner, *Die volksbiologische Forschung*, 13.

Csallner published an eight-page-long revision of Siegmund's work in 1937, tellingly entitled *Do the Poorer Peasants Really Have Less, and the Rich More Children: Are Our Communities Really Overpopulated?*[108] This brief but decisive critique of Siegmund's conceptual and methodological shortfalls was all the more important symbolically as it was published in the most prolific of Saxon newspapers, the *Siebenbürgisch-Deutsche Tageblatt*, before appearing as an offprint shortly thereafter.

Opening in the complimentary and apologetic manner so typical to Csallner's pieces on Siegmund, it states:

> We do not have many who with such eagerness and burning love for their nation have worked so tenaciously and sacrificed themselves for our continued existence as Dr. Heinrich Siegmund has done over the last decades. He has earned himself a particular distinction for having, more than any other, endeavored to teach us to see and think correctly in accordance to life's laws [*lebensgesetzlich*]. I myself owe him greatly. That is why I find it difficult, and do so only after much hesitation, to now have to publicly speak out against two of his most ardently advocated and defended views.[109]

Insisting that Siegmund's greatest achievements lay with the attempted popularization and proliferation of eugenic thought, rather than the merits of his studies, Csallner summarized his disagreement with Siegmund as follows: "More than anything else we must—no, [we must] not overcome a nonexistent overpopulation—but work towards creating overpopulation, but not with mentally and spiritually inferior people, but an overpopulation of the better and the best, so that they may finally defend and expand our *Lebensraum* and pass a surplus of German, and always German people on to our Saxon cities."[110] Highlighting the need to create a population surplus prior to any drastic expansion of the Saxon *Lebensraum* may well explain Csallner's disregard for Siegmund's Land Preservation Society, which had so actively battled to raise funds for territorial acquisitions.

While Siegmund's main focus had been with expansion, Csallner preferred a program geared towards the internal re-homogenization and the expulsion of "ethnic others." One of the clearest examples of this is offered

108 Csallner, *Haben wirklich die armen Bauern.*

109 Ibid., 1.

110 Ibid., 8.

by a later discussion of one of Siegmund's most important publications, his 1910 "The Saxon Nation's Sense of Space."[111] While Csallner agreed with his mentor's view that the Saxons lacked a clear and driven understanding of the ideological dimension of the *Lebensraum* issue, and of its practical influence on their nation's development, he repeatedly undermined Siegmund's insistence on the immediate need to expand as quickly as possible:

> He [Siegmund], and with him others, did not realize that the time for such an expansion had not yet come, that we must first retake and de-foreignize [*entfremden*] what has remained of our Saxon lands piece by piece, before we can and must begin spreading out in a similarly planned way. We must not do so in a manner that, by jumping too far ahead, creates mortally endangered or even already lost outposts—but by advancing gradually, always carefully ensuring that these pushes are added to by acquiring the land in between, in order to ensure a quick spatial reconnection with the nation as well. Siegmund has regrettably not highlighted this enough, and called for an expansion too unequivocally.[112]

Having disagreed with Siegmund so radically, Csallner turned his attention to picking apart the methodological flaws underlying Siegmund's censuses. The data that informed Siegmund's early work on the interplay between *Lebensraum* and fertility stemmed from a census undertaken in 1910. Siegmund had collected data on 30,858 families through questionnaires sent to local parish priests, and focused on correlating information on the wealth and size of 12,213, six- to twenty-year-old "fertile" marriages. Csallner was to repeat this census a quarter century later, with his revised questionnaire aptly illustrating the extent of Siegmund's methodological errors. For one, Csallner shifted his focus from families to pupils enrolled in the fifth to the seventh grades, which he thought more accurately reflected "complete" families unlikely to have more children. Most importantly though, Csallner criticized Siegmund's decision to equate wealth with property size. Because an acre of land in one location need not necessarily be equal to another, and because peasants were likely to also derive income from other means, such as renting out machinery, Csallner asked

111 Siegmund, "Der Raumsinn." Another example of Siegmund's studies which employed the same methodology and arrived at the same conclusions is Siegmund, "Der drohende Volkstod."

112 Csallner, *Die volksbiologische Forschung*, 28.

for the children's parents to be classified as "very poor," "poor," "medium," "wealthy," or "very wealthy" with regards to their local context.[113] Considering Csallner's insistence that wealth implies talent, that poverty implies inability, and that both traits are inherited rather than environmental in nature, it is actually surprising to find Siegmund retort that Csallner had neglected to account for the inferiority of these poor families that he proved to have more children than their wealthy counterparts. Writing in his own journal, the *Evangelischer Fürsorger,* Siegmund countered in the article "Long-Term Effects" ("Fernwirkungen") with the peculiar argument that merely degenerate poor peasants would have large families:

> The thoughtful peasant with little land [*Zwergwirt*] will have few children, as he lacks the very land without which they cannot subsist. [...] Not every peasant has the same understanding of the lack of land [*Raumenge*]. The indifferent, not-forward-looking, mentally deficient peasant—who might have become a *Zwergwirt* precisely because of these deficiencies—is the one who tends to have many children. This circumstance masks the *Zwergwirt*'s purposeful lack of children, and falsifies the statistical results by disfavoring the explanation that family planning is a consequence of lacking *Lebensraum*.[114]

This insistence on a lack of *Lebensraum* as the central cause for declining fertility, and the portrayal of Csallner's statistical analysis as subverted by irrational poor peasants, unaware of their objective need to limit their family size, are typical of Siegmund's reluctance to adapt his eugenic discourse to the demographic realities that Csallner so tirelessly tabulated and decried in the 1930s.

Common criticisms

Csallner, too, was bound to elicit criticism, considering his uncompromising views on nature over nurture; the two examples put forth by his fellow eugenicist Eckhard Hügel and the pedagogue Bernhard Capesius serve to illustrate this well.

113 Ibid., 28–30.
114 Siegmund, "Fernwirkungen," 3.

Eckhard Hügel, himself the author of numerous articles on the merits and necessity of racial hygiene, and head of the Research Institute's Office for Racial Studies after 1941,[115] refused to accept one of the fundamental theses underlying Csallner's eugenic discourse: the inheritability of talent. Instead, Hügel argued that Saxon settlements were historically prone to perpetual intermarriage (one of his interests lay with incest), thereby blurring any such hereditary distinction. Rather, Hügel insisted on a class struggle among the Saxons, in which the poor were suppressed by the rich: "In a discussion, he [Hügel] insisted that our villages, as a result of the high degree of communal intermixing, displayed no fundamental hereditary difference regarding talent between the poor and the rich, between the noble and the lowly; that the poor were simply too suppressed and exploited by the rich to climb the economic ladder; and that their school and lifetime achievements did not essentially overlap with one another—without me being able to liberate him from his errors."[116]

Hügel, in a review of Paul Collmer's book on the Saxon church's welfare system,[117] raised a second commonly encountered criticism, namely that of Csallner's methodology, when he complained that the author had "trusted too much in the finality of statistical studies by overlooking that Father Alfred Csallner's views on fertility as a selective process [. . .]—due to the manner in which the statistics were collected and evaluated—are not beyond question."[118]

A similar combination of complaints was also raised by Bernhard Capesius, director of the confessional boys' high school (*Knabenlyzeum*) in Bucharest. The extent of his disdain for the theses put forward in Csallner's article "The Greatest Threat" is evident, if nothing else, in his posing the question as to whether in fact Csallner's theories were not themselves "The Greatest Threat?"[119] While lauding Csallner's determination and meticulous approach to statistical research, Capesius criticized his absolute

115 Of which, in particular, see Hügel, "Möglichkeiten und Aufgaben"; Hügel, "Rassenforschung und Volksgruppe"; and Hügel, "Rassenforschung und ärztlicher Beruf."

116 Csallner, *Die volksbiologische Forschung*, 88.

117 Collmer, *Fürsorge als völkische Selbstbehauptung.* Interestingly enough, Csallner himself thought the book to be very good, apart from Collmer's overestimation of the scope and impact of Siegmund's reforms, which Csallner deemed to have lagged considerably behind what Siegmund should have been able to accomplish.

118 Hügel, review of *Fürsorge als völkische Selbstbehauptung*, 239–40.

119 Capesius, "Die grösste Gefahr?"

"faith" in statistical infallibility, and his insistence on heredity as "dangerous *Irrlichter*." Unsurprisingly, Capesius was quick to emphasize how his own pedagogical experience clearly refuted Csallner's assertion that scholarly talent was dictated by heredity. Perhaps the most damaging attack on the eugenic population policies fostered by Csallner's views is provided by the statement that:

> As understandable as it may be that a man with such faith in numbers [...] cannot see the pulpit as his actual calling, we also find it worrying to award such national significance to these works. [...] Our future [...] will not depend on relative comparisons, but on our absolute numerical size and on individuals' absolute achievements. And towards this end, may it be economically or spiritually, each and every one should be educated and called upon equally. [...] The blessing though is given or withheld by God—and not by statistics.[120]

Csallner was clearly incensed about the criticisms of his eugenic vision, and complained that "it has repeatedly come to pass that during a conversation, or even at a later point, objections have been raised that—at least with a part of my audience lacking the appropriate expertise to instantly notice the errors of these ridiculous complaints—have hemmed the influence of my observations. My theories have also been doubted and contradicted in print, and even I have myself been attacked."[121]

So, in response to these two indicative assaults on his definition of Saxon degeneration, Csallner published four retaliatory, or "assisting studies" (*Hilfsarbeiten*) as he termed them, between 1935 and 1938:

1. "A False Comfort"[122] set out to defend the thesis that inherited talent was teetering towards extinction, and to assess the dissatisfactory "nature" of replenishing talent from the lower classes because, quite simply, "untalented parents cannot have highly talented children," and because he did not "prefer to believe in the miracle that once more talent of the high and highest sort had simply appeared out of nowhere";[123]

120 Quoted in Csallner, *Die volksbiologische Forschung*, 88.
121 Ibid.
122 Csallner, *Ein falscher Trost.*
123 Ibid., 3, 5.

2. "On the School Grades Achieved by Parents, Children, and Their Siblings"[124] endeavored to prove not only the statistical feasibility of deducing and quantifying the child's talent from his/her parents' achievements, but that the inverse was also the case;
3. "*Paarungssiebung* among Transylvanian Saxon Peasants"[125] argued that peasants tended to choose spouses with similar scholarly achievements and otherwise similar talents—while insisting that both the "better" and "inferior" (who reproduce more rapidly) tended to group together, creating something of a natural hierarchy of families; and finally
4. "The School Achievements and Lifetime Achievements of Transylvanian Saxon Peasants"[126] argued that precisely this correlation was irrefutable, as superior talents result in greater economic success and social standing—defined, for example, by honorary positions held—and that these most successful members of a given community unavoidably bore more talented children with better grades than their peers.

The mighty pen

If Csallner was always going to struggle to convince his most ardent critics, he also repeatedly attempted to popularize his views through poetry, plays, and prose. Interestingly enough, and occasioned by Csallner's ninety-fifth birthday in 1990, the *Siebenbürgische Zeitung* published a brief laudatio of his life's achievements entitled "Working towards the Nation's Health," which included the peculiar (and carefully phrased) assertion that "particularly Alfred Csallner's literary works clarify what he actually wanted: to educate his countrymen, to encourage them to act in a particular manner towards the benefit of the Transylvanian Saxon nation—for that is how he understood its *Gemeinschaft*."[127]

124 Csallner, "Über Schulzeugnisse."
125 Csallner, "Paarungssiebung."
126 Csallner, "Schul- und Lebensleistungen."
127 Wa[gner], "Wirken zum Wohle der Gemeinschaft."

Csallner produced an unexpectedly wide and varied wealth of poems, short stories,[128] and even a play, *For the Future's Sake.*[129] According to his sister Elfriede, Csallner also published under the pseudonym Friedrich Nösner.[130] Nonetheless, the treasure trove of fiction Csallner has produced under his own name is remarkable in more ways than one. Not only largely autobiographical, Csallner's literary outputs are remarkably concise and illustrative abstracts of his overall eugenic and race-hygienic discourse.

One the most instructive insights Csallner has offered on his perception of the dysgenic crisis and its cure is found in a poem published in *Selbsthilfe* in 1931, entitled "Eight Hundred Years . . . !" In it, Csallner lamented that while only a bare few had truly understood the gravity of their situation, the majority blissfully insisted on a sort of immunity acquired over the 800 years of Saxon struggle in Transylvania. Before coming to the grim conclusion that it would not be his and his companions' fault were the Saxons to become extinct, Csallner furnished the reader with some intriguing sound bites, and two short, illustrative excerpts are very much worth translating (even if at the cost of the poem's rhyme):[131]

> And in vain are all the cautions
> that we stop to foolishly throw away, with both hands,
> the nation's wealth as we have until now,
> to no longer destroy budding life,
> to no longer allow ourselves to be thinned out by emigration,
> because foreign nations squeeze themselves into the gaps,
> constricting our space year on year,
> tolerated by us, even called upon by us,
> ennobled and elevated by us onto the highest rungs,
> also enabled by us to advance with certainty of victory,
> conquering and growing at our expense,

128 Of these, Csallner's 1980 collection of short stories is particularly interesting: Csallner, *Der Baruch*. Also see two illustrative short pieces published in *Der Aufbau* in 1934: Csallner, "Des Esels Trost"; and Csallner, "Das Märchen von Königen und Köhlersleuten." Unsurprisingly, the moral of the stories is that one's hereditary qualities are innate and immutable, regardless of education and upbringing.

129 Csallner, *Um der Zukunft Willen*.

130 Csallner, Elfriede. "'Es ist fast nichts übrig geblieben . . .'"

131 Csallner, "Achthundert Jahre . . .!" Also see Csallner, "Du und Dein Volk"; and his poem "Higher Upwards!" published in one of Csallner's final assignments as director of the LSS, namely the "hereditary passport" (*Ahnenpass*) issued by the *Volksgemeinschaft* in 1940: Csallner, *Ahnenpass*, 41.

while we rest and rust on wilting laurels,
we who are missing the chance to save our future
by dreaming of great deeds past
and think our history stands
sentinel over our continued existence.
[...]
However much this realization may hurt,
he who hadn't scorned and continued to study
why we have shrunk, lost, and retreated—
he would find a knowledge, one unmatched
in worth, and would jubilantly hold in his hands
the means to turn our destiny around:
to regenerate, allow us to grow and
no longer miss an opportunity
to furthermore expand our life's space
so that the foreign will to exterminate would crumble
faced with our will to stay live,
to green, to blossom, and to bear fruits.

IV. Fritz Fabritius's Self-Help, from "Building Society" to Rebuilding Society

The emergence, evolution, and empowerment of Saxon fascism in the wake of the First World War, and its eventual *Gleichschaltung* with the German model at the hands of Andreas Schmidt in 1940, have attracted considerable historiographic attention and controversy. But despite the plethora of publications defining the Self-Help and its subsequent permutations as the carbon copy of a glorified German National Socialist ideal at one end of the scale, or as an ultimately democratic force at the other, the movement's eugenic core somehow seems to remain an undercurrent at best, a blind import at worst. Perhaps precisely due to the specificities of this historiographic climate, the existing literature has largely neglected to locate the Self-Help within wider debates on the interwar period's political and eugenic discourses.

So, what was the Self-Help, and what did it want? In general terms, the Self-Help was a remarkably self-absorbed ideological synthesis of "nationalism" and "socialism," striving towards a eugenic national regeneration, a triadic agenda that eventually adopted National Socialism, but that was nonetheless distinct from it.

Ideological import or indigenous movement?

The single most divisive issue to bedevil the Self-Help's evaluation is whether the Self-Help was an indigenous movement, the result of a process of integration and adaptation, or a wholesale import of German National Socialism. As shall be argued throughout, the Self-Help was effectively all three of these, but at different stages in its development, which can be identified as: the "origins," 1922–32; the stage of "politicization," 1932–40; and the *Gleichschaltung*, 1940–44. The reality that most university-educated Saxons studied in either Germany or Austria (and, naturally, Budapest prior to 1919) undeniably infers a higher-than-usual degree of knowledge transfer, so the key question must lie with evaluating the selective appropriation of ideological thought and practical policies rather than an either/or approach. No ideology connected to the international community evolved in isolation, but as is the case with the Saxon eugenics and Saxon fascism, an ideology can very well be the local permutation of larger processes.

The Self-Help began as an indigenous discourse preoccupied with local issues and local means of addressing the needs of the urban poor in Hermannstadt.[132] No matter how many times the "legendary" encounter between Fabritius and Hitler in 1922 is recounted, to advance this as evidence for a eureka moment following which Fabritius returned to Transylvania with a new ideology lock, stock, and barrel is simply ahistorical (as is the allusion to the Self-Help's use of the swastika, as Siegmund's BSV had already sported it on their letterheads in 1914). This is not only because there currently is no concrete proof that such a meeting ever took place (in addition to the somewhat divergent accounts thereof), but because its significance does not lay with the movement's early development at all; much rather, it was a means of political legitimization after 1930. In fact, the move-

132 That the Self-Help, at first, offered little more than an ideologically branded form of economic aid that appealed to the poor considerably more than to the educated classes is evident in its repeatedly stated aim to attract more "intellectuals." See, for example, Orend, "Der Nationalsozialismus und die 'Intellektuellen.'"

ment's organ, the *Selbsthilfe*, only began printing articles borrowed from the *Völkische Beobachter* in 1929, and articles on National Socialist ideology are not found anywhere in its pages before 1930 (while it should also be noted that those found in later years were almost always Hitler-centric biographies rather than conceptual debates).

The first programmatic piece to discuss National Socialism to any meaningful depth was Karl Hermann Theil's 1930 three-part article on "Adolf Hitler and National Socialism." Typically for the plethora of similar articles that were to follow, it amounted to little more than an idealized biography of Hitler (divided into the four sections "Hitler the man," "Hitler the critic," "The origins and rise of National Socialism," and "The National Socialist state") that concluded "National Socialism's ideological foundation is formed by Hitler's critical insights, developed during his *Lehrjahre* in Vienna and during the war."[133] Not only does 1930 appear rather late for a movement that would insist to have been National Socialist from the very outset, but it is characteristic of the movement's Hitler-centricity and otherwise wanting engagement with National Socialism as a political ideology.[134]

The junction at which the Self-Help's repeatedly professed sympathies for Hitler's National Socialism become crucial is the proposition that the Self-Help had not actually assimilated National Socialism, but imported it wholesale. While even some secondary sources expound this thesis, it is problematic not only because the Self-Help at no point seems to have debated National Socialist ideological content, but because it repeatedly insisted that while they pursued similar aims to National Socialism, the means were necessarily different given that the Saxons were an ethnic minority. Again in Karl Hermann Theil's words, "our goal must be the transfer and adaptation of these ideas to our nation's benefit and good. We certainly don't want to set up SA and SS formations, we certainly do not want to copy and imitate with-

133 Theil, "Adolf Hitler und der Nationalsozialismus."

134 So, for example, G., "Nationalsozialismus"; W., "Kühl bis ans Herz hinan," 2, which insisted that National Socialism was the healing process of deadly ill Germany—indeed, its "rebirth,"—without actually defining what that meant; Wr., "Nationalsozialismus und christliche Religion," which maintained that its supposed anticlericalism was an international conspiracy, and that National Socialism actually represented "positive Christianity"; "Der Nationalsozialismus (Grundsätze und Bewegung)," which followed the common template of presenting Hitler as a First World War veteran, come symbol of Germany's reawakening; and, finally, S., "Schalgworte?," one of the few examples that differentiated between German National Socialism as a political party and its value as an "organic" ideology—the latter of which the Self-Help identified with.

out criticism, but we want to create the spirit."[135] What he understood by "the spirit" of National Socialism, Theil clarified in a subsequent piece: "Adolf Hitler's idea thundered over us like a tempest, shaking us and demanding action, the awakening from a deep national lethargy. All that was healthy among us raised its head from the slumber of national will-lessness, and who felt the call of the blood stepped into the newly created front delighted to take part and be able to act."[136] It was precisely this combination of revolutionary zeal, the sense of a new beginning, and the promise of a genuine sense of eternal community that defined National Socialism's ideological appeal[137]—one certainly shared by wider segments of the population.

It is within the context of the Self-Help's rebranding as the National Socialist Self-Help Movement of the Germans in Romania—i.e., when it founded a political party, in 1932, subsequent to the movement's voracious expansion and radicalization—that one can reasonably question the indigenous quality of its ideological discourse. But even at this point, there are no proven grounds to believe that the NSDR was a wholesale import. Much rather, and all the more so coming from Wilhelm Schunn, who had been opposed to the movement's active involvement with party politics, the Self-Help adopted German National Socialism as a franchise, as the casing for its existing (if perpetually evolving) ideological tenets:

> As with the motherland where Adolf Hitler's battles had been spiritually prepared [*geistig vorbereitet*] by mostly unknown individual champions, so too the soil amongst Romania's Germans had been prepared by a movement for renewal reaching back into a time even before the Great War. This rooted movement for renewal, though, had not yet conclusively decided what it wanted. It was only when Adolf Hitler's plans became known here, and when the new—more German and stronger—type of man created through him began marching in front of the world, that it, too, found its destination [*Endziel*] and the clear spiritual form it has today.[138]

135 Theil, "Nationalsozialismus und Selbsthilfe," 1. A further example insisting that as an ethnic minority the Saxons simply could not transplant the German model is offered in A. K., "Unsere Politik." Although, of course, the Self-Help did in fact set up a SA (Self-Help Work Teams), later SAM, that summer.

136 Theil, "Der Nationalsozialismus und das Deutschtum in Rumänien," 1.

137 Dörr, "Warum wir Nationalsozialisten sind?"; Pomarius, "Warum wir Nationalsozialisten sein müssen."

138 Schunn, Weg und Feinde der NEDR, 3. That said, Schunn was outraged to realize that an unnamed hand had "revised" the passages of his book that voiced his opposition to the transformation. In a written statement, Schunn insisted: "In 1932, Mr. Fabritius, influenced by a group of radicals from Braşov, abandoned

In other words, German National Socialism offered the Self-Help something of a template with which to format its ideology, not the ideology itself, even if the two displayed apparent similarities regarding their perceptions of a nation in perpetual decline, and the shared yearning for a "new man" to be born of its ashes. However, this format was adapted piecemeal, and particularly so with regards to its practical suitability to the reality of Saxon political options. This is a fundamental point, all the more so as the Saxons, as an ethnic minority, had to operate within certain conceptual and practical limits. Perhaps the single best example of this is offered by the much referred to dichotomy between the movement's authoritarian (dictatorial even) insistence upon enforcing the *Führerprinzip*, and it advocating an essentially liberal-democratic electoral policy. With the decision to enter the fray by founding the NSDR as a political party, the Self-Help had reached a watershed, one that necessitated an exhaustive overhaul of Saxon electoral practices, were it ever to become politically empowered. But a clear differentiation must be made between the two targets that can happily coexist: the first is the presumption of the ideology's underlying need for a hierarchical chain of command and the individual's absolute subordination to the nation's needs in the absence of an internal monopoly on violence (the basic precondition for a state); the other offered the only viable route to political power. Saxon fascism could not possibly advocate totalitarianism in principle, as given its minority status doing so would have fundamentally undermined its basic ability to operate. Therefore it is not surprising that Otto Fritz Jickeli, one of the movement's key figures, who also opposed its transformation into a party, and who repeatedly insisted that while National Socialism and the Self-Help shared similar goals they must pursue different means, asserted that "the advantages stemming from a liberal democratic understanding of statehood have had beneficial results for minorities. The exaggeration of nationalist thought in this world means a worsening of the minorities' position."[139]

---

the nonpolitical path of the old Self-Help and created a political movement. In opposition to this change in direction, I refused to assume any function in the political movement established in 1932." Enraged by the revisions, Schunn continued, "I was disgusted and I showed my displeasure vehemently." See ZAEV, Bestand Schunn, no. 10, pp. 1–2. The anonymous "editor" even sent a short apologetic note to Schunn, claiming: "This entire mess is not my fault. I was given the order to revise those sections that allowed for the conclusion that there had been differing opinions about leaving the nonpolitical path." ZAEV, Bestand Schunn, no. 7, p. 1.

139 Jickeli, *Unser Weg*, 4.

The synthesis of political ideology with national economy

Even if other European fascist movements failed to develop viable economic policies, politics and economics are anything but mutually exclusive. On the contrary, the Self-Help was built around the interconnected conceptual triad of politics, eugenics, and the economy. This was reflected by its organizational structure as early as July 1922, when the Self-Help subsumed three distinct spheres of activity: first, the original Saxon Workers' Book Club (Deutsch-sächsische Arbeiter-Lesezirkel), espousing an urban ideology geared towards a poor urban audience; second, the Settler Group (Siedlungsgruppe), founded upon the centrality of the *Lebensraum* motif and associated racial, social, and public hygiene agendas; and third, the Economic Group (Wirschaftsgruppe) that perpetually criticized modern capitalism as a "foreign," degenerative force corrupting the Saxon nation in spirit (the quest for profits) and body (the banking system's lack of altruistic funding for social projects).[140]

In other words, the Self-Help was by its very nature a political ideology—albeit one that unto 1932 believed it could achieve more through lobby work and private projects than in the form of a political party (an anyway alien concept given Saxon history). But given the absence of a political arena even remotely resembling that of a nation-state in form and function, it is entirely logical that the Self-Help concentrated on those domains it could impact—namely, the reorientation of economic means towards internal colonization and ethnic homogeneity.

What is more, the matter was indeed conclusively resolved with the 1931 publication of a leaflet detailing the aim and ambitions of its economic activities. Written by Schunn, who was to repeatedly refer back to it as proof that the Self-Help had at no point considered economic policies to be its chief purpose, the first paragraph insists:

> What is the Self-Help? The Self-Help is *Kampforganisation.* The movement thinks it a failure that affects us all if our nation does not give each of its members worthy thereof sufficient *Lebensraum* and an idealistic *Weltansschauung* on his way. It [the movement] wants to conquer our entire nation in order to renew it

140 See "Volksgenossen."

> ideologically, economically, and racially. It fights for the principle that only those in our midst who work honestly, live simply, and are noble in mind [*hochgesinnt*] and deed are honored. The Self-Help deals with those economic concerns that others do not want to touch, and where it is hence forced to fill the gap.[141]

The Self-Help had rapidly evolved from a small reading circle advocating urban farming as a means of lifting the urban poor out of poverty to a complex political ideology by around 1929. Nonetheless, the Self-Help's approach to the economic crisis it had initially been founded to combat was always deeply political.

The primacy of eugenics

Eugenics, and by extension racial hygiene, were the Self-Help's ideological glue. Despite various and numerous secondary sources either ignoring or cursorily dismissing the existence of a Saxon eugenic discourse, it was absolutely central to the "movement for renewal's" vision of national regeneration—as is perhaps most appropriately illustrated by one of its key figures, Wilhelm Schunn, when he marvels: "Racial anthropology's achievements (this timely arrived and indispensable gift of providence towards German reconstruction) have always been discussed and revisited. They have created the solid foundation for a German idealistic and yet rooted worldview."[142] In other words, eugenics offered the key both to diagnosing the "cancer" afflicting a Saxon nation perceived as teetering on the brink of extinction, and to distilling a cure as dictated by the commandments of racial purity and ethnic re-homogenization. The Self-Help discussed eugenics long before National Socialism became a defining issue, and did so because it offered Saxon fascism a tool by which to reconfigure virtually everything from national identity to cultural and economic policies. A tool by which to build a eugenic fortress, clearly, when Fabritius asks: "Which are the foundations that guarantee any national community's continued existence? Apart from the community's typological homogeneity (racial homogeneity) [*Artgleichheit* (*Rassegleichheit*)], which is self-evidently the natural precondition

141 Schunn, *Das neue Merkblatt*, 1.

142 Schunn, *Weg und Feinde der NEDR*, 11.

for shared moral values, a shared sense of law and honor, and its communal economic policy, the availability of sufficient *Lebensraum* is the central issue from which all paths towards a healthy higher evolution lead."[143]

The Self-Help's eugenic ideology was, in essence, defined and refined by the cumulative influences exerted by Heinrich Siegmund, Alfred Csallner, and Wilhelm Schunn. Siegmund joined the movement's ranks in 1927 (in fact, he was the first member in Mediasch, and predated the formation of a local branch by five years), and Csallner's Society of Child Enthusiasts followed suit in 1929. These dates become truly interesting in the context of Siegmund's church-based Welfare Committee, which not only joined the Self-Help, but created a space in which Siegmund, Csallner, and Fabritius collaborated. While Siegmund's overall influence on the movement's eugenic discourse was implicit, Csallner in particular made the Self-Help his new ideological home with the creation of his Race Office in 1932.

"Saxon, Saxon, above all else"?

The Self-Help, while thoroughly pan-German, was peculiarly divorced from virtually anything non-Saxon. Apart from the occasional brief news bulletins, the *Selbsthilfe* very rarely discussed political/fascist movements or events outside of Germany and Austria (not including the occasional rant against the ills of Bolshevik Russia). A very telling omission testifying to the Self-Help's relative self-absorption is the absence of any discussion of the rise and empowerment of Fascist Italy. One of the few useful examples available, though, comes in the form of a defensive article written in response to a blistering criticism of an "open evening" (*Sprechabend*) organized by the Self-Help in Hermannstadt. The article offers a rather favorable review, arguing that Italian art and culture had never had as many options as they had then: "We see under the grip of Italian party membership [*Kandare der Parteizugehörigkeit*] a blossoming second Italy, we see drained swamps and freshly planted slopes saved from karstification. We see, instead of a weak and corrupt gypsy nation, a new powerful Italian nation emerge."[144]

143 Fabritius, "Heimaterde," 1. For a full translation of Fabritius's article, see Georgescu, "Transylvanian Saxon Eugenics."

144 S., "Sie bitten ums Wort," 1. Naturally, he also discusses Bolshevism, defined as a subhuman ideology lashing out against all that is holy.

The degree to which the Self-Help almost exclusively focused on Saxon affairs (at least to any meaningful depth) is illustrated by its relationship with the Romanian state, which it tended to accept as a force of nature that ravaged the Saxon nation but was otherwise removed from the realities of the day-to-day struggle for survival with the Romanian nation that increasingly appropriated the Saxon economy and *Lebensraum*. The Self-Help hence professed a rather peculiar, if all the more telling relationship (which it defined as "foreign policy") with the Romanian state, in that despite the incessantly derided policies coming from Bucharest, its pan-German ideology necessitated loyalty. The Self-Help's longing to embark upon a second nation-building project through the crystallization of a new, "800,000-soul nation" of all Germans in Romania automatically condemned any irredentist ambitions. Indeed, the Self-Help, in word and print alike, was awash with celebrations of the creation of Greater Romania, thought to have offered the multitude of German settler colonies which it subsumed an unprecedented opportunity. Despite the fundamentally different trajectories that the Banat, Bessarabia, Bukovina, and Dobrudja "colonies" had taken (including the disjointed concentrations of Germans in the Old Kingdom), and the difficulties inherent in tying together such culturally, religiously, economically, and politically divergent groups, the Self-Help rejoiced at the new opportunities their unification under one political entity had created.[145] To that end, the movement strove to dominate not only the Saxon National Council, but also the Association of Germans in Romania, which it had deemed a catastrophic failure. The Self-Help led them in 1933 and 1935, respectively, before disintegrating into civil war between the movement's radical and moderate factions.

v. Wilhelm Schunn's National Neighborhoods and Honorary Gifts

Wilhelm Schunn (1888–1966) was a key figure in the Self-Help's formulation of a clear, coherent ideology during its rise to power, and instrumental to the movement's attempts at translating dogma into practice after 1933. He also had a significant influence on Csallner, who frequently mentioned their

145 See, for example, Jickeli, *Unser Weg zur Erneuerung*, 7. He also makes a rather intriguing reference to these "tribes" being racially more diverse than those in Germany. Ibid., 8. See also Jickeli, "Kundgebung der NEDR in Kronstadt"; Jickeli, "Die Kundgebung der NEDR im Burzenland."

close friendship, ideological kinship, and collaborations, hailing Schunn as "the genius regenerator [*Erneuerer*] and propagator of the neighborhood system."[146] Schunn is nevertheless regularly overlooked by secondary literature. His peculiar absence from Saxon historiography is surprising considering he wrote the movement's twenty-five-point program (which for its part also served as the NSDR's political program), the Saxon 1933 national program, and the 1935 "National Program of All Germans in Romania," in addition to overseeing the wide range of departments and offices affiliated with his "regenerated national neighborhoods," and which were subsumed under his wider jurisdiction as the Commissioner of the Nation's Organic Constitution in 1935.

Born on March 27, 1888, the twelfth of thirteen children, Schunn lost his mother age six—a childhood trauma that suffuses a thoroughly intriguing CV he submitted to the church in support of his application for a parish post in September 1951.[147] Towards expounding his religious credentials, Schunn reflects on having joined the puritan White Cross temperance movement aged eleven or twelve: "I wanted to keep my body holy and free of all stains, which I have held until I was grown up." He also details his suffering at the hands of an unnamed lung disease that ravished most of his adolescence, which he claims to have contracted from a sick teacher with whom he had lived for a year, prior to being cured at a sanatorium near Jena. Schunn had studied theology, mathematics, and physics in Jena, Leipzig, and Klausenburg before returning to Hermannstadt to work as a teacher.

Having spent the First World War as an army chaplain, Schunn returned to his teaching post in Hermannstadt and got married in 1923. Oddly enough, especially given his eugenic views, Schunn's marriage produced no children. By the late 1920s, he became increasingly "melancholic" (a trait he seems to have shared with his youngest sister, who committed suicide in

146 Csallner, *Die Volksbiologische Forschung*, 4. Schunn reciprocated the praise in a letter addressed to Csallner on his seventieth birthday, in which he wrote: "I know of no one whose life had been as determined and dedicated to the existential battle of our desperately struggling little nation as yours." Csallner, *Zur Verleihung des siebenbürgisch-sächsischen Kulturpreises 1982*, 3.

147 ZAEV, Bestand Schunn, no. 1, pp. 1–4. Interestingly enough, this short autobiography includes a passage in which Schunn claims God had allowed his mother to visit the "living" in the form of an old lady: "I knew that we were neither allowed to stop or talk to each other. Because God was watching. He sometimes allowed her, in the form of an old lady, to see her boy. But the silence and the distance of death were not to be violated, or else her return would no longer have been permissible [*gestatet*]." Ibid., 2.

1929),[148] a sense of crisis only alleviated by plunging himself into reinventing the Saxon neighborhoods model.

Despite these claims to a profound religiosity, Schunn had stirred controversy in April 1932, when outraged parents complained to the city priest (*Stadtpfarrer*) of Hermannstadt, Friedrich Müller, that Schunn in his mathematics class had told their daughters, first, that "Mass was actually only for *Hottentotten* and niggers, while it should be a service to mankind [*Dienst am Menschen*]," and second, that "Jesus never taught to love thy neighbor of his own accord—his followers asked him to. He himself would have never said that. Jesus didn't create the educational system [*Lehrgebäude*], his disciples did."[149] In other words, Schunn sought religiosity—a "sacred canopy"—in and through society, not the Bible. Ultimately, he found it with the neighborhoods' re-enchantment of public time and space. Csallner was to play a vital role in distilling and implementing their eugenic policies, especially with regards to the honorary gifts; the training of Hermannstadt's thirty-nine individual neighborhoods' genealogy officers (*Amtswalter für Sippenwesen*) in January 1937;[150] and by disseminating his interpretation of the interplay between heredity, race, and *Lebensraum* in numerous public lectures.[151]

At the very heart of the Self-Help's bid to create a eugenic fortress we find Wilhelm Schunn's national neighborhoods. The executive to the fascist-dominated National Community of Germans in Romania's legislative, Schunn's organization was absolutely pivotal to the fascist bid to exert an increasingly totalizing degree of control over its body politic, to regenerate the nation's "spirit" through the re-enchantment of public time and space, and most importantly here, the neighborhoods were seminal to any feasible attempt at translating eugenic population policies into practice. However, they are entirely absent from the available secondary literature—surprisingly so, considering the astonishingly comprehensive treasure trove of archival material held by the Romanian National Archive in Sibiu. While the investigation offered here proposes initial conclusions on the neighborhoods' specifically eugenic and race-hygienic discourses, as evidenced by

148 Ibid., 4.
149 ZAEV, Bestand Schunn, no. 5, p. 2.
150 NA Sibiu, CGR, D.XIV.841, p. 1. See also NA Sibiu, CGR, D.XIII.B.612, p. 193.
151 NA Sibiu, CGR, D.XIII.B.477, p. 28.

their ideological imperatives and methodological means, this, surely, can be but the beginning of historiography's interest in the neighborhood system. All the more so as it frequently resembles an Interior Ministry populated by twenty-two various individual (and frequently overlapping) offices and institutions above and beyond those discussed here.

In short, Schunn's national neighborhoods served three essential and interdependent purposes.

The totalitarian executive

The neighborhood system of micromanaging virtually every facet of the German minorities' political, cultural, economic, and family life was reinvented by Schunn based on the Saxon historical template banned by the Hungarian state as subversive in 1871. Building on this sense of historic legitimacy, the neighborhoods reemerged thoroughly suffused with the indigenous brand of fascist dogma espoused by the Self-Help movement, which had undergone a most remarkable transformation from book club to empowered regime within a decade of stepping into the public domain. This cross-pollination of the revered ancient with utopian visions of national resurrection, so characteristic of any fascist project, is nowhere as apparent as in the wide range of nation-(re)building tasks with which Schunn entrusted his neighborhoods. Designed as practically self-sufficient local self-help collectives subsuming some one hundred Saxon families in cities and fifty to sixty in rural communities, each neighborhood was in charge of, amongst others: the collection or requisition of fiscal and material donations; the distribution of gifts on festive occasions such as weddings, births, and funerals; coordinating the communal help given to members building or rebuilding homesteads; ensuring that property sales did not diminish a Saxon *Lebensraum* deemed to be in constant peril; discouraging mixed marriages; preventing ethnic "others" from undermining Saxon businesses and industries by acting as an informal employment office and tailoring the supply of Saxon labor to Saxon economic needs; guaranteeing a high standard of medical provision; and, ultimately, enforcing their members' ideological indoctrination in the name of national salvation. In short, they were omnipresent in everyday life from the cradle to the grave.

Re-enchanting time and space

These closed community support networks furnished fascist politics with a vital tool toward legislating policy decisions at the local level, but the fascist bid to build a eugenic fortress required a far more subtle means of ensuring their actual execution, above and beyond the *Führerprinzip* and the threat of internal exclusion. Fusing ideological imperatives with structural necessity, Schunn's neighborhoods sought to create an "organic" family suffused with a sense of shared sacred spaces and symbols. If nothing else, this appropriation of idealized historic traditions, and their infusion with the fascist imagery of health and spirituality towards a redefined, alternative modernity, is astonishing. Of these, perhaps the singularly most revealing example is provided by the "neighborhood chest" that while utterly superfluous, was endowed with a deeply religious function. Be they old relics or new variants, these chests were ceremoniously "consecrated" (*Ladenweihe*) by the neighborhood that placed them at the heart of any of the numerous, equally symbol-laden public gatherings.

The primacy of eugenics

The quest for totalitarian politics (despite all assurances to the contrary) toward a nation reinvented found its ultimate expression in the primacy of eugenics. Amongst the myriad of social-hygiene and public-health agendas, the singularly most imposing example of how Saxon eugenics materialized on the ground comes in the form of the neighborhood policy of awarding honorary gifts to celebrate the birth of the fourth or more child of a "racially valuable" family—a policy that was deeply indebted to Schunn and Alfred Csallner's close friendship and collaboration. This gift came in the form of a one-off payment of a staggering 20,000 lei for the fourth, and a further 10,000 lei for any subsequent children, when the overwhelming majority of recipients had a monthly income around the 2,000–3,000 lei mark. However, there were strings attached: the money had to be invested in "goods of lasting value," in accordance with a preapproved spending plan, and potential recipients had to pass a race-hygienic evaluation of their "hereditary fitness" to ensure that these funds encouraged a qualitative rather than purely quantitative population increase. While it is important to note that at the moment there is little more than circumstantial evidence that these race-hygienic exams were actually carried out, the gifts' very purpose and conditional clauses testify to a fundamental shift in the Saxon approach to identity. Not only did it embrace a distinctly biological framework, but it also resorted to a truly draconian internal dynamic, where Saxons failing to live up to an imaginary racial and moral ideal were exiled from an ethnically exclusive community. In other words, the honorary gift demonstrates the significance and perceived utility of a eugenically defined vision of national regeneration to a Saxon nation perceived as teetering on the brink of dissolution, which responded by revolutionizing the ideological characterization, anthropological classification, and political instrumentalization of its own racially framed identity.

CHAPTER II

ASSESSING THE DYSGENIC CRISIS: KEY CONCEPTS AND THESES IN ALFRED CSALLNER'S DEFINITION OF SAXON DEGENERATION

Csallner featured heavily in the elaboration of the core conceptual issues of Saxon eugenics: demography and differential fertility; *Lebensraum* and the loss thereof; and the host of socioeconomic factors aggravating both, including emigration, mixed marriages, and the economic losses incurred by employing non-Saxons, or by wasting national funds on alcohol and tobacco. But how did these themes evolve and unfold during the brief but intense burst of activity between the mid-1920s and 1940? In short, Csallner was concerned that inferior parts of the population had begun to increasingly outweigh the racially best endowed, and decried the continued degeneration of hereditary substance that this entailed. In his view, the root evil threatening the Saxon nation's status as *Kulturvolk,* and even its very existence, was not merely an overall decline in fertility, but the internal disparity between the birthrates exhibited by various professions and the extent to which these reflected their hereditary worth. Emigration patterns more widely, and in particular the socioeconomic strata from which the emigrants originated were of similar interest, as Csallner lamented the loss of talented individuals (and, of course, the loss of potential future offspring). Yet one of the most sinister side effects of emigration, according to Csallner, was that it created an excess of unmarried Saxon women left with little choice other than to resign themselves to solitude or else marry non-Saxons. To Csallner's mind, mixed marriages were not merely a race-anthropological aberration, but a symptom of the Saxon national spirit's accelerating decline into oblivion. Interwoven into all of these discourses one finds the omnipresent question of *Lebensraum*, the continuous loss of which not only marked the logical outcome of a proliferating "two-child system," but acted as its catalyst. Paired with the Saxon tendency to employ "foreign" day laborers and servants, Csallner contended that the Saxons had effectively paid the Romanian "invaders" the very wages they needed to buy their way into Saxon settlements. Further, he argued that this resulted in a downward spiral racing towards the eventual dissolution of the Saxon *Lebensraum* the nation so desperately needed in order to rejuvenate and repopulate its national body.

1. The Lost Children: Family Planning and the Demographic Collapse

Csallner loved numbers. His entire system of thought assumed the quantifiability of virtually any aspect of Saxon economic and spiritual life—historical processes and their outcomes could be defined in terms of statistical analysis. Csallner accumulated vast databases of numbers in the course of his career, with which he aspired to unequivocally prove that declining birthrates lay at the heart of the Saxon dysgenic crisis. Before delving into the hereditary "worth" of the various Saxon professions and their offspring as Csallner understood them, the overall demographic constant in Csallner's eugenic discourse needs to be evaluated here. Prior to engaging with his discussion of class-specific "quality," in what follows I discuss Csallner's theories on the nation's overall "quantity" in light of a spreading "two-child system" that not merely cheated the Saxon nation out of its offspring numerically, but produced inherently inferior children.

The study of demographic trends had been the earliest of Csallner's eugenic interests, and dominated many of his early publications. Virtually all his related interests, be they rural employment patterns or marital or drinking habits, answered to and fed into his lifelong battle to encourage the proliferation of large, racially valuable families. His insistence on the fundamental necessity of primarily positive eugenic measures constituted the beating heart of his regenerative agenda. Therefore it is little surprising to find Csallner demonizing the proliferation of small families as a curse, a disease, a degeneration of not only the Saxons' hereditary worth, but their spirit; he writes that birth control "poisons our souls over and over again and numbs our hearts: it spoils our nation's soul [*Geist*], its values, and way of life."[1]

In one of his earliest articles, the merely eight-page-long "Our Losses at the Hands of Family Planning,"[2] Csallner set out to calculate, tabulate, and illustrate the extent to which the Nösnerland had been cheated of its children. In a manner truly typical of his reverence for and methodological ease with numbers, he compared the region's demographic growth between 1765 and 1813 to the data available for 1905. To then derive the discrepancy of people actually "lost," Csallner produced a projection of the Nösnerland's prospective population had fertility rates not begun to decline. Using this "ideal" rate of population growth, Csallner subsequently insisted on the much earlier dates by which each of the twenty-two rural communities could—or, rather, should—have reached their current size, had they continued to grow as quickly as in the eighteenth century. Indeed, rather than the mere 14,768, the population would have grown to 50,100 by 1905—a size equivalent to his projection for the year 1822—had they reproduced at a continuous rate. In other words, population growth had been stunted by approximately eighty-three years, and the nation had been denied 35,340 lives it so desperately needed in the ongoing ethnic struggle for *Lebensraum*.[3]

Csallner introduced his article with the following, almost apocalyptic assault on birth control and family planning as the most poignant cause and consequence of societal indifference and decline:

> It is a disease, a degeneration of the heart that thinks itself caring for wanting to leave its children with as large and undivided an inheritance as possible, and that nonetheless is raw and sinful enough to prevent from being born all those other children that could come, or—if they do want to come—to even murder them in the womb.
>
> And it is a disease of the mind [*Verstand*] that deems itself so smart and crafty but still refuses to acknowledge how it damages the entire nation, as well as his descendants.
>
> But now we know more, see things more clearly than we used to. And there may be none amongst us who don't realize that we are damaging ourselves, who don't wish things would change. But there are too few of those who

1 Csallner, "Unsere Verluste." The same phrase is also found in Csallner, "Unsere einzigen Kinder," 2.

2 Csallner, "Unsere Verluste."

3 Ibid., 6.

> won't sluggishly and stupidly leave it to others to act, of those who are willing to break with the foolish and sinful traditions, who—by their deeds—help to rid our nation of its curse. And that is why it is a disease, the degeneration also of our will.[4]

It is precisely at this junction of national will and national want that Csallner raises his eugenic agenda, his moral finger, and insists that had the population grown as it clearly could have, "far fewer non-Saxons would live in our midst, [and they] would not be as dangerous economically, culturally, and politically as they actually are today."[5]

Projecting these processes from the Nösnerland onto the larger Saxon stage, Csallner reached similarly disquieting conclusions, as seen in his graph on "The Increasing Lack of Transylvanian Saxon Children". Surveying Siebenbürgen's total population (the data for which came from the 1937 census conducted by his LSBS), Csallner's attention focused on the period 1914–33. More precisely, he endeavored to tabulate the reproductive cost of the First World War, and concluded that had the Saxons continued at the same fertility rate as in the immediate pre-1914 years, the overall population would have grown by 121,054 children by 1933. Instead, Csallner lamented the shortfall of 16,716 children, or the equivalent population of the entire Schenker church district.

In other words, the numbers are endowed with meaning by virtue of the amount of *Lebensraum* they could have populated/secured. So, for example, in his 1928 "Our Only Children" Csallner further expanded upon the idea that declining Saxon fertility had initially created the need and opportunity for ethnic others to infiltrate their towns and villages (the ethnic distributions of which he then proceeds to tabulate), to learn and benefit from their Saxon neighbors, and to do so to the dire detriment of Saxon power and influence.[6] Csallner then turned his attention to the manner in which the existing populace seemingly wasted their finances on alcohol and tobacco instead of investing in their national advancement, as well as the "great economic burden embodied by the large number of inferior individuals."[7]

4 Ibid., 2–3.

5 Ibid., 5.

6 For example, Csallner, "Unsere einzigen Kinder," 1n1.

7 Csallner, *Betrachtungen zur Lage*, 8.

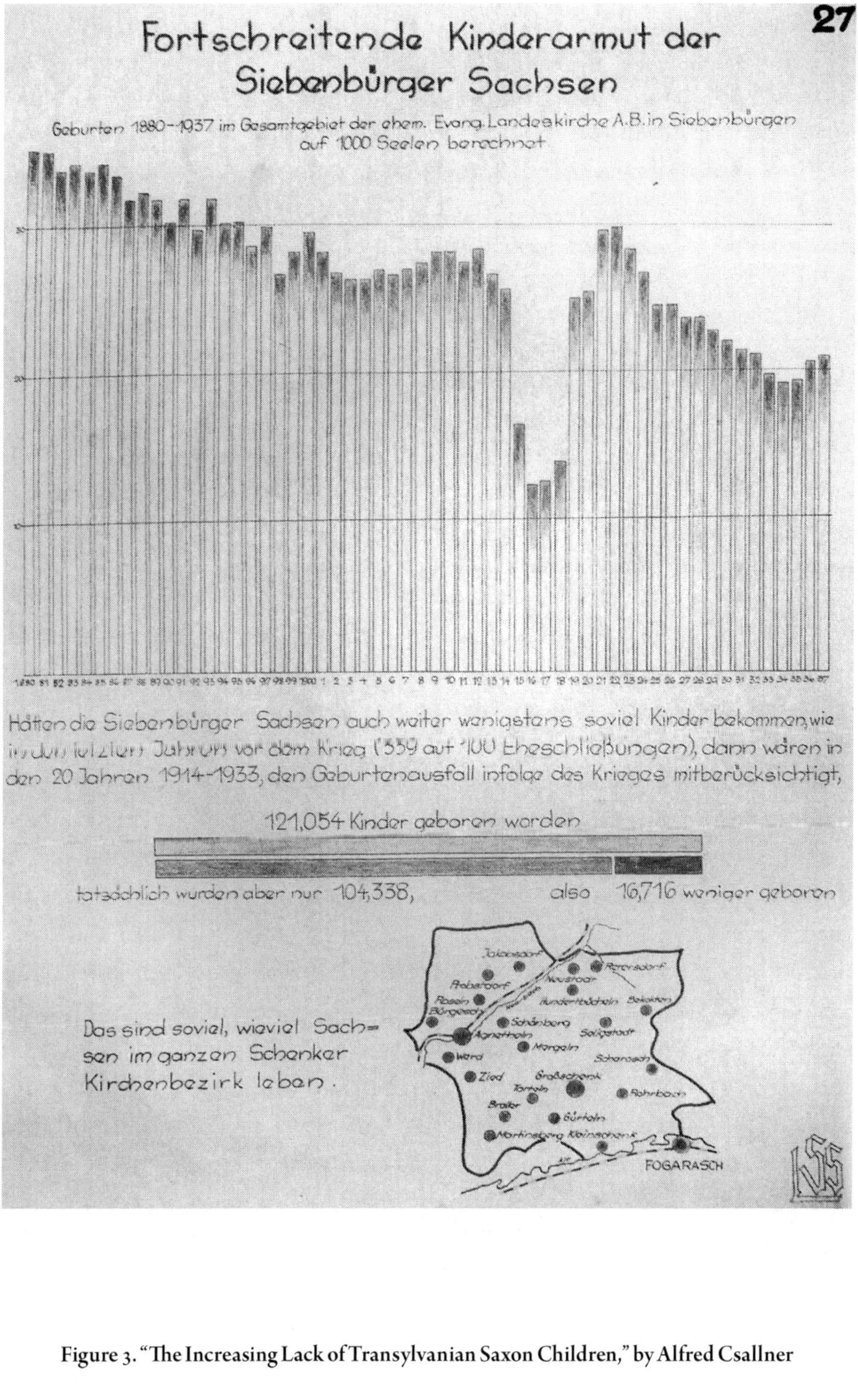

Figure 3. "The Increasing Lack of Transylvanian Saxon Children," by Alfred Csallner

Apart from the numerical impact birth control was having on the competitive strength of the Saxon nation, Csallner was adamant that the offspring of single- or two-child families were effectively inferior to those born into large families. The following passage, taken from his 1936 article on "National Biological Studies" published in the *Medizinische Zeitschrift*, reveals his entrenched and almost bitter disdain of single children in particular, whom he vilified as a growing degenerative influence:

> But this heavy loss in fertility is itself the consequence of a certain wimpishness [*Verweichlichung*] and sissiness [*Wehleidigkeit*], of a lack of fighting spirit, of a certain smugness and egoistic attitude. And looping back they [single children] further weaken [*verweichlichen*] our nation, its strength to act and willingness to sacrifice, diminishing its internal cohesion and will to defend and extend its property—so that the nation, in any case, cannot fight the battle for continued existence in the way that it could, according to its nature. If one even acknowledges that the lack or complete absence of siblings has a negative influence on character development, one will hardly want to deny how strong the influence that they [single children] exert on the views of the national community must be, when one is told how many people we have with only few siblings or without any.[8]

Csallner's remarkable twelve-page-long pamphlet *Our Only Children*, published in 1928, provides a particularly tantalizing grand summary of the detrimental effects small families supposedly exerted on the very substance of their individual members, and how these, by extension, would pass on their moral and physical inferiority to future generations. Leaning heavily on and quoting extensively from the research of the Viennese pediatrician and psychoanalyst Dr. Josef Friedjung, Csallner underscored that his findings revealed a meagre thirteen out of one hundred single children to be "normal," the rest becoming little more than a "pitiful type of man." What is more, a third of these one hundred children were deemed physically inferior due to their perceived tendency towards eating disorders, and because they were more likely to be pale, weak, and unlikely to recover from common illnesses as quickly as a "normal" child would.[9]

8 Csallner, "Die volksbiologische Lage," 356.
9 Csallner, "Unsere einzigen Kinder," 2–3.

Having thus alluded to the physical inferiority of single children, Csallner turned his attention to their spiritual deficiencies—and to the work of the pediatrician Dr. Neter—when characterizing them as "cowardly, not independent, clumsy, vain, with narcissistic tendencies, and both selfish and unsociable."[10] Csallner set his ideal minimum requirement at four children, if any were to truly live up to their full potential, as families with merely two children offered no verifiable change to the behavioral patterns he had expounded. Effectively, the larger the family the better, as Csallner's rather awkward (not to mention difficult to prove) line of reasoning insisted "both history and daily experience show us clearly enough that customarily, children from large families turn out better and achieve more—while the offspring of single- and two-child families so frequently degenerate, or at least fail to become that which, if they had had more siblings, could have become."[11]

It is remarkable, if not merely indicative of Csallner's own lack of medical knowledge, that he at no point in this (or, for that matter, any other) exposé of the inherent inferiority of single children discussed the apparent dichotomy: while Csallner, on the one hand, so avidly and rabidly insisted that all forms of "talent" are inherited rather than nurtured, he also contended that these inferior children passed on their environmentally acquired deficiencies to their children, perpetually degrading the Saxon nation in its entirety. A good example is the following thesis: "Where we find already degenerated parents, where single- or two-child families are particularly common, they exert an exceptionally harmful influence on the respective community, indeed on the entire nation."[12]

Then, in a rather interesting shift of focus, Csallner proceeds to explain a further degenerative byproduct of family planning, but in relation to infant mortality and wedding patterns. Offering the example of twenty-two rural Nösnerland communities, Csallner explains that of the 1,000 brides wed between 1906 and 1910, 648 were under twenty years old, and 198 not even sixteen. Considering that, as he believed, the first and second born are the most inferior of children, and that most families limit their size to either number, then these anyway inadequate children are additionally burdened

10 Ibid., 4.
11 Ibid., 5.
12 Ibid.

by being reared by a mother herself too young and weak to do so properly.[13] But as the "healthier" children she could have born in later years were never conceived, this process produced a perpetual downward spiral, where the few children actually born would themselves become deficient parents to deficient children at too young an age: "Now consider: In most cases the frequently weakest, least resistant first children stay alive and reproduce, often much too young themselves, while the other children that could come and would want to come—born to the now older and stronger mothers—would mostly be themselves stronger and healthier, and would in turn have stronger and healthier children; but these are barred from entering life hundreds and thousands of times: must that not have a damaging impact on our nation?"[14]

So if according to Csallner the overall decline in Saxon fertility had first (indeed, both in terms of push and pull) opened the gates to the subsequent flood of ethnic others swamping their cities and settlements economically, chipping away their *Lebensraum*'s cohesion territorially, and undermining the Saxon nation racially, then the dysgenic crisis was only deepened, and it accentuated the creation of a cycle where degenerate children become degenerate parents to perpetually degenerate children.

II. The Quality Question: The Nation's Hereditarily "Best" under Threat of Extinction

As his insistence on the inherent inferiority of small families illustrates, Csallner understood the spread of single- and two-child families as both the cause and consequence of degeneration. But if this, so to speak, represents the "macro-level" of Saxon demographic trends, then Csallner was even more concerned with variations at the "micro-level"—the differential fertility rates exhibited by those socioeconomic classes he deemed superior versus those he deemed inferior in talent and worth. The conclusions he derived from these surveys diagnosed a process of accelerated decline, even forecasting the eventual disappearance of racial and cultural substance, beyond the purely numeric loss in generic human capital.

13 Two related studies, as far as age is concerned, focused on the number and "nature" of children born in wedlock but conceived prior to marriage: Csallner, "Vorehelich gezeugte Kinder in Roseln"; and Csallner, "Vorehelich gezeugte Kinder in einigen Gemeinden des Schenker Kirchnbezirks."

14 Csallner, "Unsere einzigen Kinder," 8.

In the fall of 1926, Csallner begun researching and publishing again on the reproductive trends of what he had classified as the "untalented," "talented," and "most talented" sections of the Saxon population as early as 1919, but the bulk of his work in this field was published much later, in the early to mid-1930s. Unwavering in his insistence on "nature over nurture," Csallner became obsessed with the detrimental effects that birth control was having amongst the nation's hereditarily best endowed. Focusing on the parents' wealth, social standing, and school grades as his variables, Csallner surveyed the scholastic achievements of some 19,000 school children between the fifth and seventh grades for indicators of inherited intelligence and talent. Guiding his evaluation was the entrenched premise that the wealthy must necessarily be talented, that the recipients of high social honors were endowed by nature with a superior character, and that untalented parents could not possibly bear talented children. With that and the varying reproductive cycles of these three categories in mind, Csallner set about calculating the long-term repercussions. He quickly concluded that the ongoing loss of highly valuable hereditary stock would continue beyond repair unless immediate action was taken. While the overall decline in Saxon fertility represented the macro-level of what Csallner perceived as a growing dysgenic, indeed existential crisis, the sociobiological (*gesellschaftbiologische*) micro-level of class- or talent-specific fertility was to encompass the bulk of his most ferociously defended views on heredity: "The rapid decline in our hereditary industriousness has a more disastrous impact than anything else. This is because the better and best talented—who, precisely due to their better talent, haven risen higher economically, socially, and mentally [*geistig*]—emigrate much more frequently, remain unmarried more commonly, above all else though, have less children and consequently decrease in number, or at least proportionally lag behind the large and mostly lesser valuable sections of our nation."[15]

Csallner believed that the superior sections of society, which had initially given birth to the *Kulturvolk* but were now falling victim to emigration and family planning, were increasingly being outnumbered by their untalented and inferior, but more fertile counterparts. While this trend led to a clearly quantifiable weakening of national strength and resolve, Csallner

15 Csallner, "Die volksbiologische Lage," 356–57.

also feared that the swelling tide of racially inferior traits would ultimately rob the Saxons of not merely their literary output, but their very ability to sustain the economic growth which rendered them the "most advanced" of Transylvania's nations: "But this means that if this fateful pattern continues, the [nation's] bearers—the rich and very rich families—and all the traits that empower their economic success and protection of acquired riches, will become rarer, while those inferior traits of the reproductively stronger poor and poorest, which hinder economic success, will become more frequent. [. . .] Economic industriousness, though, is a precondition for our ability to defend our culture and nationhood."[16]

The urgency with which Csallner began publishing on this particular degenerative trend was reinforced by his insistence on a cyclical pattern underlying the rise and fall of nations, of civilizations. While his 1939 article "On the Regeneration and Degeneration of Nations"[17] (which was to quote extensively from Lothrop Stoddard's *Der Kulturumsturz*) is one of his later pieces, it accurately sets out the doomsday scenario so intrinsic to his entire line of reasoning. It was precisely such a change in demographic processes that forced a high civilization to "sink into pitiful meaningless."[18]

Csallner, who held a strained and selective understanding of hereditary biology, showed a pronounced dislike of what he thought were "verifiably" inferior sections of society, which he percieved as "retarding agents":

> The inferior, finally, are incapable to accept and to learn: that is to say they stay behind and can but retard, or even destroy.
> But this discrepancy in a people's inherent ability to grasp culture [*Kulturfähigkeit*] has nothing to do with chance, it is conditioned by heredity. Whatever they are worth, they have received it from their parents, and they will pass it on to their children.[19]

So, in short, Csallner believed nations were born by a process in which hereditary traits of various qualities would inexorably condense into mutually exclusive reproductive groups. He argued that these would compete with each other demographically, and that the more talented portions of

16 Csallner, "Die grösste Gefahr," 251.
17 Csallner, "Von Aufartung und Entartung."
18 Ibid., 13.
19 Ibid., 13–14.

society would initially win the upper hand, as the ranks of the innately inferior would be decimated by the poverty brought about by their very inability to keep up with cultural progress. But perhaps more drastic still, Csallner considered their gradual decimation because of low fertility and short life expectancy as a cleansing process of paramount cultural utility: "That these weakest stay behind, though, yields particularly beneficial results for the community, the tribe, the nation. Because they are too weak spiritually, physically, and morally to keep up with culture, they live in worse conditions and can thus reproduce slower, and die quicker—declining in numbers and shrinking away frequently into complete extinction—and the tribe, the nation, is continuously cleansed of its useless offspring."[20]

Under the banner of "social biology" (*Gesellschaftsbiologie*), Csallner began a host of surveys to assess the extent to which this particularly degenerating "side effect" of family planning was undermining individual settlements as organic wholes, and individual trades and professions he deemed particularly valuable. While many, such as the survey of the Saxon medical profession, were never brought to completion, three of the sociobiological studies that were published clearly stand out. The first, which focused on the traders and artisans in Bistritz, was also one of his earliest publications, and Csallner was delighted to repeatedly point out it was the very first such study of a profession to be carried out amongst the Saxons.[21] The two other studies, published in 1932 and 1934, respectively, investigated the dysgenic crisis threatening the academic classes in Mediasch[22] and Bistritz.[23] (Although Csallner had initially also gathered the data for Fogarasch, he concluded it had "too few" academics to merit an independent evaluation.)

With the support offered by local government, Csallner began collating data on tradesmen and artisans in Bistritz as early as the summer of 1919. But enjoying little success in completing his survey of the Saxon and non-Saxon economies and of their overlap, he was to wait until the town's Academic Union (Akademische Vereinigung Bistritz) conducted a related census in

20 Ibid., 15.

21 Csallner, "Statistisches über den Bistritzer sächsischen Gewerbestand." In 1929, Csallner was to apply the same approach to qualifying the extent to which declining fertility rates undermined Saxon economic survival to the case study of Agnetheln and its tradesmen, in a much abbreviated form. See Csallner, "Beispiele für die wachsende Kinderbeschränkung."

22 Csallner, *Der Mediascher deutsche Akademikerstand.*

23 Csallner, *Der Bistritzer deutsche Akademikerstand.*

1924 and handed him their data.[24] Although both sets were incomplete, their synthesis allowed Csallner to produce his first sociobiological study of a profession's economic standing and marriage and fertility patterns, and crucially, to assess what became of its children. This study was by no means a purely descriptive analysis of Saxon economic degradation, but created the statistical basis for population policies geared towards an increasingly centralized form of career and job-market management to combat the growing number of non-Saxon employees and apprentices.

Csallner was alarmed to find Saxon traders increasingly reliant on their workshops (pointing to a decrease in income derived from the secondary, agricultural sources), and utterly dismayed by the "dangerous" number of Romanian and Hungarian apprentices in Saxon trades. When Csallner found that of the 153 businesses 85 had no apprentices, and that of the 103 existing posts only 53 were filled by Saxons (while 25 were Romanians, and a further 23 Hungarians), he concluded two things. First, he reasoned that although the Saxon tradesmen statistically bore enough children to replenish themselves, "strong social prejudices on both sides have meant that too many tradesmen's sons are striving, in a downright pathological manner, towards other, mostly academic professions—without there being a sufficiently strong countermovement from other classes to replenish it [the trades]." Csallner's second, considerably more alarmist conclusion, was that this self-inflicted shortfall of Saxon apprentices was being compensated with the very "ethnic others" with whom the Saxons existed in a permanent struggle for economic survival: "And so we ourselves, with this erroneous behavior, encourage the foreign ethnicities' advancement into virtually all domains of business. We ourselves lure, train, and strengthen foreign apprentices and journeymen in the economic struggle that they, once independent, will lead against us."[25]

The two sociobiological studies Csallner published on the impact of the dysgenic crisis on the microcosm of academic classes in Mediasch (1932) and Bistritz (1934) employed a largely identical methodology and conceptual framework, and were seminal to his understanding of how differential fertility rates contributed to the decrease and eventual extinction of what he

24 Csallner, *Die volksbiologische Forschung*, 93; and Csallner, "Statistisches über den Bistritzer sächsischen Gewerbestand," 1.

25 Csallner, "Statistisches über den Bistritzer sächsischen Gewerbestand," 11.

deemed to be the nation's best hereditary traits and qualities. The themes, or agents of degeneration, threading through both of these studies are a particularly valuable reflection of Csallner's overall eugenic agenda, in that they bemoan the continuous ascendancy of non-Saxon competition; the high degree of Saxon academic emigration; the Saxon academics' tendency to "import" wives from Germany and Austria, thereby not only changing the Saxon racial composition but creating a "surplus" of Saxon women for foreigners to pray upon; and the late age at which they start their (inevitably small) families, thereby depriving the nation of its most valuable offspring.

Of the few commendable trends Csallner was able to discern, he was particularly adamant about the manner in which internal migrations acted as proof for increasingly strengthened blood ties (*Blutsbande*), melding the various Saxon settlement areas into one organic whole guided by a communal destiny (*Schicksalsgemeinschaft*). This focus on urban "catchment areas" was of particular value to Csallner's propagandistic efforts, in that it allowed him to underscore the extent to which all Saxon settlements had to pull together in one unified eugenic effort to combat degeneration, and no village could remain passive while a beleaguered neighbor was struggling for survival. Csallner was particularly pleased with his map "Fortification of Blood Ties and a Communal Destiny by Internal Migrations," which detailed the places of origin for Bistritz's academics, Mühlbach's craftsmen, and the priests, teachers, and professors to have found their way to Hermannstadt. As the map showed a wide geographic distribution, Csallner "expected it to have a good educational impact," by demonstrating that even the smallest of rural communities nonetheless made important contributions to larger cities, and that those "big and strong towns must, for their own survival's sake, help the small and weak communities in their struggle for survival, particularly with regards to sustaining the church and schools."[26]

Returning to his concern with racial integrity and the impact of the Saxon academics' tendency to search for wives abroad, along with German academics immigrating to Saxon lands, Csallner points to the "racial changes" these imply.[27] The internal migrations also illustrate two of Csallner's subsequent theses. First is his thesis that the fates of the various Saxon

26 Csallner, *Meine wissenschaftlichen Arbeiten* [1975], 16.

27 Csallner, *Der Bistritzer deutsche Akademikerstand*, 4. Csallner actually quoted this from his first study on Mediasch's academics.

settlements are intertwined—the geographical spread of Bistritz's "catchment area" underlines the extent to which the Saxon nation was not a conglomeration of isolated splinters, but an organic whole. The second thesis is that the increasingly desperate need of the academic caste to replenish itself from outside its own ranks is a sign of its "sickness," brought about by insufficient offspring who, if they are to study, either emigrate or pick up professions different from those of their parents.[28] That these processes point to the Saxon nation's overall racial degeneration is illustrated by Csallner's conclusion: "Overall, the better and best talented grow from other castes into the academic one, which makes its decline all the more regrettable. It not only stunts our overall growth, but simultaneously leads to our nation's continuous impoverishment of the good and best hereditary traits, a decline of our racial worth."[29]

For his first, 1932 sociobiological study on Mediasch, Csallner had collected data on 155 working or retired academics in 1926 (including their families, this survey subsumed about 5,000 people, or about a tenth of the total Saxon population). Thus, working with about twice the volume of data he was to compile on Bistritz, Csallner arrived at largely the same conclusions, criticizing an alleged Jewish dominance over the local economic life, along with the gradual expulsion of Saxons from public offices by the Romanian state's ambitions of internal colonization. Csallner emphasizes again the manner in which internal migrations perpetually mold and strengthen the Saxon nation as an organic whole, when he writes: "We are practically all related to all, and these blood ties are tied and retied ever tighter. And we awake to this blood union as a communal destiny when we realize how large the catchment area of, for example, Mediasch is, and that it is not possible for Mediasch not to care whether Weilau retains its Protestant school, or whether Gürteln's Saxonness withers or strengthens."[30]

Moving from the micro- to the macro-level, in 1935 Csallner published what was going to be his most substantive (and lengthy) study of Saxon hereditary worth, with the telling title *"We Want to Remain What We Are"—Really?* In short, it reiterated the familiar line that the discrepancy in fertility

28 Ibid., 5–6.

29 Ibid., 8.

30 Csallner, *Der Mediascher deutsche Akademikerstand*, 5–6.

between the highly talented and the untalented families inexorably led to a "reduction also of inner worth, of the capacity for culture to such an extent that we will be unable to sustain ourselves as a nation—and that should this come to pass, it would not really be a pity if we were to end."[31] There is a note of resignation in Csallner's tone, one he underlines with the statement that "I have—at times alone, at times together with others—tried to stop this collapse, but both the masses and the majority of our national leaders and subleaders have responded with disbelief and indifference, indeed even with resistance and hostility."[32]

Beginning with the customary lamentations of the continuous loss of *Lebensraum* to a Romanian nation either buying land or stealing it through the state's rural reforms, Csallner underscored the degenerative effects implied by the declining fertility rates amongst the Saxon nation's various castes, and again insisted that hereditarily "better" parents invariably produce "better," more "productive" offspring, and then ventured to forecast the long-term repercussions should these trends continue unabated. He comes to the conclusion that the data demonstrated that: "The percentage of highly valuable families, in 150 years, will fall from 34 to 24 percent, the proportion of medium-valuable will grow from 40 to 41 percent, so hardly change, but the proportion of inferior families will grow from 26 to 35 percent. This group of inferior families, though, can mostly only aggravate, complicate, and destroy the life of our nation, but must be dragged along by the national community."[33] These proportional changes implied that in 150 years 100 "highly valuable" families would have to contend with not 76 but 146 inferior families vying for resources and power.

This apocalyptic vision of Saxon substance being degraded at the hands of hereditarily inferior social strata is particularly apparent in one of Csallner's many graphs, "The Decline in Our Industriousness as a Result of Differential Fertility," compiled for Kronstadt. Visualizing his understanding of how the hereditary makeup of the city's German population would fundamentally change in favor of the least desirable sociobiological classes, Csallner believed that the proportion of the "very good" classes would fall by some two thirds by 2070, that of the "good" by about half, while the "aver-

31 Csallner, "Mer welle bleiwen," 1.

32 Ibid.

33 Ibid., 10–11.

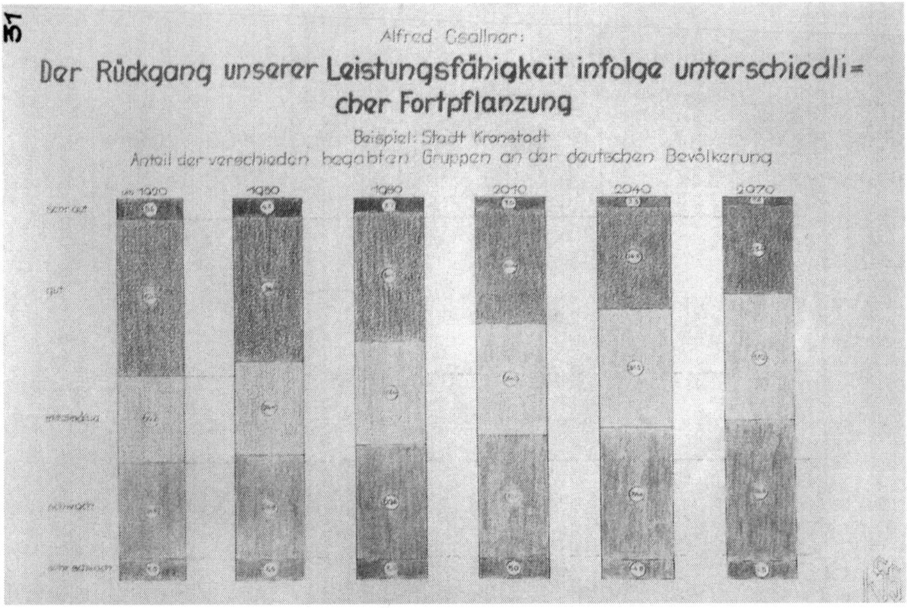

Figure 4. Alfred Csallner: The Decline in Our Industriousness due to Differential Fertility (Kronstadt).

age" group would rise by about a third, that of the weak by about a quarter, and the proportion of the "very weak" would largely stay the same. The key point to take away from this graph is that Csallner did not prophesize that the "very weak" would explode numerically, but that they, through the infiltration of other castes, themselves too weak reproductively, would have a cumulative effect dragging down all of society's strata.

Refocusing on the town of Bistritz, Csallner retained the same pessimistic outlook exhibited above. Apart from the Saxons' continuous loss of everything material and immaterial to ethnic others, Csallner's perception of his hometown's ethnic and economic composition is almost always host to rants against a perceived Jewish dominance largely absent from most of his other case studies. The following is not merely an astonishing example thereof, but aptly illustrates the undercurrent of frustration and disappointment present throughout this piece:

> [Since 1918] the percentage of Germans has decreased further, while that of the Romanians and Jews has only increased. And the Saxon merchant and business professions have maybe shrunk faster still. One store after the other

> collapses, is dissolved, or sold. From the other direction advance unhindered—indeed, in an unfathomably short-sighted, suicidal manner, even furthered by the entirely unenlightened Saxon rural population—the conquering Jews, who open new stores, expand the old ones, buy up hitherto German businesses. And a Romanian merchant caste is already growing.[34]

Adding to these otherwise competitive processes undermining the viability of Saxon economic life in Bistritz, Csallner continued to lay the blame, at least in part, at the feet of the Romanian state, which he criticized for having "displaced" the Saxons from their administrative positions, and which he believed yearned for their economic "annihilation." Nonetheless, Csallner, as in many of his previous texts, always also blames the Saxons themselves when he exclaims that "no, if we go under—and Bistritz is particularly endangered—then it will be not because others murdered us, but because we are ourselves rotten, have not defended ourselves, and allowed others to murder us!"[35]

That said, it is little surprising that Csallner answers the question posed in *"We Want to Remain"* by stressing that the Saxons were far from "remaining what they were"—on the contrary, they were degenerating, losing their physical and mental capacities for culture and resistance, and were likely to go extinct if the nation did not awake soon to the "horrific" nature of the dysgenic crisis besieging it. It is with a certain melancholic undertone that Csallner, in light of his political views and the Third Reich's ascent to power, finds himself in 1934 envious of a Germany embarking on the very same race-hygienic crusade he had demanded for the previous twelve years:

> It is with pride, joy, and sorrowful desire that we see how Hitler and the Third Reich have begun their historically greatest deed—how they salvage the German nation from certain demise, and save it maybe even for millennia. They, of course, have a state, a German state that upholds and supports the German nation and has the power to enforce what is necessary where enlightenment cannot help and willingness is lacking. We Germans have no state of our own; and no matter how good our intentions towards our state [...], its power is still in many ways against us. And that which must be done

34 Ibid., 14.

35 Ibid., 15.

> to save our nation here, we must achieve by our own strength and in spite of the state—and having so little external power to force those of us who do not accept, who do not want to partake.[36]

III. Emigration: The Loss of Saxon Hereditary Substance

While emigration was always a prominent feature of Csallner's understanding of the dysgenic crisis, it rarely took the center stage. Much rather, it served as a demonstration of the extent to which a lacking national program to guide the supply of and demand for various professions had created a scenario in which either a critical mass of Saxon youngsters found themselves competing for the same positions, or certain socioeconomic fields were forced to resort to "ethnic others" in the absence of Saxon labor.

Csallner only published a few articles on emigration specifically. By the time he furnished a case study, in 1940, on the prevalent emigration patterns in the village of Grossalisch,[37] his tone had markedly changed from earlier publications pleading for a containment of emigration. By 1940, Csallner's thought had become more radical, isolationist, and willing to wield the eugenic axe handed to him by the increasingly institutionalized race-hygienic discourse.

In a bid to raise awareness of eugenic ideals, Csallner had begun to cluster teachers and priests into "working groups" in the early 1930s, and in 1938 assigned one of them the task of collating and evaluating all available data on emigration from their respective area. *Study on the Emigration from Villages, the Example of Grossalisch* was the product of the first set of data made available to him. It had an educational brief and was supposed to serve as a "best-practice" guide to how he expected the other working groups to fulfil their tasks.

Employing the methodology so familiar to his work by now—namely, the study of school children's report cards as barometers of their inherited talents in relation to their parents' wealth and social standing—Csallner endeavored to illustrate the extent to which the "best endowed" comprised the bulk of the émigrés, and the deteriorating effect that had on the

36 Ibid., 17.

37 Csallner, *Die Erforschung der Abwanderung vom Dorf.*

hereditary substance of the village they left behind. Furthermore, it aimed to investigate their choice of destination and to enquire how many eventually returned, in what state of mind, and importantly, with whom. A telling example of Csallner's increasingly expulsive stance towards those deemed undesirable to the Saxon community is afforded by his brief evaluation of the five Saxons from Grossalisch who emigrated to the Old Kingdom: "One returned, but with a Hungarian wife and ruined by his stay abroad. Of a fourth it is said he was also likely to return. But those returning from the Old Kingdom usually bring so much that is undesirable with them—see the mentioned case—and are themselves corrupted, that one is not necessarily sure whether to actually wish they had better not come back."[38]

That said, Csallner quite unexpectedly arrives at the unusual conclusion that Grossalisch had actually benefited from emigration, because its least valuable members had departed, leaving the village more vibrant than it had been before: "If greater industriousness amongst our peasants truly leads to greater wealth, then the remaining Germans of Grossalisch—while giving away many an individual—have not seen their hereditary worth diminished. Rather, it has increased because the resident Germans are, on average, wealthier and thus better talented and more industrious than those who left and have remained outside."[39] However, Csallner challenges his own conclusion with the second of his preferred indicators of inherited quality—honorary positions. And while he argues that seen from that angle the ancestral substance of Grossalisch was indeed damaged by emigration, he decides that wealth was, when competing with honorary positions, the more appropriate of the two yardsticks.

IV. Mixed Marriages: The End of Racial Distinctiveness

Of the host of themes Csallner grouped together as the motors of a perceived Saxon racial degeneration, the study of mixed marriages was to become one of his central lines of enquiry from the early 1930s onwards. In short, Csallner believed that their increasing frequency (and, importantly, the ethnic group with which they were most common) reflected the moral ills afflict-

38 Ibid., 5.
39 Ibid., 8.

ing the Saxon nation, and a weakening of the Saxon spiritual resolve—and eventually physical ability—to arrest an ongoing loss of hereditary nobles and racial cohesion. Unsurprisingly then, Csallner was adamant that mixed marriages constituted one of the greatest of existential threats, and not merely in terms of the numerical loss of Saxons drawn to their spouse's culture, but because the infusion of Saxon blood into any of its ethnic neighbors would, by nature, strengthen and elevate these—their very competitors—in the struggle for survival:

> Most have come to realize that the Saxon nation, faced with ever increasing instances of mixed marriages, loses more than it gains. But most have mainly understood this trend in terms of the manner in which the children born into mixed marriages are frequently immersed in the "foreign" nation's culture. But only a few have understood the repercussions for our, and their, hereditary substance overall, and racial constitution in particular. [... It will not be necessary to prove that] the infusion of foreign hereditary traits undermines our fitness and industriousness, while the surrounding nations profit from the infusion of German blood—and that thus strengthening our opponents in the struggle for survival poses a danger to us.[40]

So, while adamant that any mixed marriage amounted to an obvious loss in terms of sheer numbers, and that children resulting from such marriages are commonly immersed in the "others' culture," Csallner's greatest worry was their impact on the Saxons' racial constitution. His assessment of the consequences of mixing hereditary traits amounted to a doomsday scenario when he stressed how foreign nations were ennobled by the introduction of Saxon blood, while it was degenerated by the inferiority introduced to the Saxon nature through mixed marriages.

From a popularizing point of view, Csallner continued to stress that any mixed marriages inevitably undermined popular opposition to them, a euphemizing process that must inevitably lead to the blurring of the Saxons' ethnic distinctiveness and, ipso facto, of their ability to exist as a nation—a conclusion dramatically illustrated by the calculation that it would take ca. 300 years (or around ten generations) before the distant descendant of an individual born of a mixed marriage, under the condition that all descen-

40 Csallner, "Mischehen Teil I," 226–27.

dants consistently married Saxons from there on in, would be almost fully reclaimed by the Saxon nation, with a ratio of 1/1024 non-Saxon to Saxon hereditary traits.[41] In other words, one of Csallner's key concerns lay not only with the immediate impact that mixing "foreign" with Saxon racial traits had on either constitution, but with their longevity, the inordinate amount of time necessary to expunge them again from the national body. The more frequent mixed marriages were to become, argued Csallner, the more difficult it would be to quarantine those (disadvantageous) hereditary traits, and the more they would seep into the overall Saxon national body, undermining both its innate nature and its viability to exist as a distinct nation:

> What is more, it cannot be claimed that there has only been one mixed marriage and none subsequently, as more and more are being formed. And so all links in the national chain will, the longer this practice occurs, come to inherit more traits foreign to their nation, from an increasingly large number of ancestors. And that is why we, no matter how German we feel ourselves to be, will become less truly German, and simultaneously more Romanian, Hungarian, Czech, or otherwise foreign Germans. That is how the mixed marriage threatens us, precisely because it has become so common, and appears to become even more so, endangering not merely our particular nature, but our survival as a nation.[42]

These underlying fears of racial assimilation (in contrast to a purely linguistic integration) formed the core of Csallner's opposition to mixed marriages, and embodied the keystone of his aspiration to design and build an ethnically homogenous "eugenic fortress."

Csallner attributed the very occurrence of mixed marriages and their increasingly rapid proliferation to a wide variety of factors. He was concerned with the ongoing emigration of young men unable to find employment locally, and with the extent to which a vulnerable surplus of Saxon women had been created by the Saxon academics' tendency to marry German or Austrian women. Csallner's assessment of the interplay between this biopolitical "supply and demand" oscillated between sympathy and accusation for the bereft Saxon women "forced to either remain unmarried or marry foreign men."

41 Ibid., 227.
42 Ibid.

But he also frequently interpreted their crisis as one of their own making, by insisting that Saxon women had come to set an unreasonably high economic bar, and that consequently any Saxon man shied away, "as he could never satisfy these demands, and deems it possible to find a satisfactorily, modest, and homely women only amongst a foreign nation."[43] Either way, Csallner believed that the root cause of mixed marriages was the Saxon nation's utter failure to preach and enforce its cohesion ideologically, to awaken in its members the inner yearning to work towards the nation rather than personal happiness: "Mixed marriages have become more frequent because we have not understood, or at least not actively attempted, to raise a nation proud of its inherited nature, to teach it the responsibility and necessity of willfully subordinating oneself to the needs of the whole [*Gemeinschaft*]."[44]

While Csallner was adamant about the Saxon nation's lacking will to tackle the problem posed by the growing influx of ethnic others either dogmatically or practically, he also deemed mixed marriages to be an example of the detrimental socioeconomic processes and inter-national realities at work. As the region's various nations had begun approaching something of a sociocultural and economic equilibrium, he was all the more vocal in his insistence that "in particular, parts of the Romanian nation have consciously set out to marry away our girls, and with them, thus, our property."[45] This rather interesting assertion seemed all the more real to Csallner, in that he understood it not only as a feature of the historic battle for *Lebensraum*, but as a direct consequence of the Romanian state's proactive attempt to supplant Saxon (and Hungarian) civil servants with Romanian ones. In a more subdued tone, he claimed that "since 1919 so many thousands of young Romanians have climbed the career ladder that they are unable to find enough women to marry of their own ethnicity and of the class into which they have risen, and are forced to grab into other nations, particularly the German, if they don't want to marry beneath their status."[46]

The bulk of Csallner's work on mixed marriages is found in a three-part article published in both the *Auslandsdeutsche Volksforschung*[47] and the

43 Csallner, "Zur Frage der Mischehe," 9.

44 Csallner, "Mischehen Teil I," 227. See also Csallner, "Zur Frage der Mischehe," 8.

45 Csallner, "Zur Frage der Mischehe," 8.

46 Ibid., 8–9.

47 Csallner, "Mischehen Teil I," "Mischehen Teil II," and "Mischehen Teil III."

Saxon *Medizinische Zeitschrift* in 1937/38.[48] Csallner proceeds to examine, where possible, the marriage patterns of the Saxon, Romanian, Hungarian, and Jewish populations, and their degree of intermarriage in the towns Sächsisch-Regen, Bistritz, Schässburg, and Hermannstadt, mostly for the period 1895–1937. The key variables Csallner employs in the subsequent tabulation and excavation of the statistical data (largely originating from his aborted 1937 survey) relate the total number of mixed marriages to the respective gender marrying a "foreigner," the groom's profession, the bride's age at the time of marriage, and, of course, divorce rates. In other words, Csallner set out to study the relative "quantity" of mixed marriages in terms of their historic trends, the individual socioeconomic castes least and most likely to marry foreign spouses, and, as Csallner was in any case convinced that mixed marriages could do little other than fail, to evidence the strife they inherently brought into any home. But his focus on the bride's age is of particular interest, as to Csallner's mind it indicated the underlying reason why Saxon women intermarry: if they were younger than the average Saxon bride, that implied a wanting national(ist) mindset; if they were older, they had probably waited for a Saxon suitor as long as possible before finding themselves forced to marry anyone rather than stay unwed.

Csallner's overall conclusions presume that the number of mixed marriages was not only unexpectedly high, but rising, while Saxons increasingly tended to marry Romanians rather than Hungarians (whom they had largely preferred prior to Transylvania's union with Romania). The reason for this shift lay, according to Csallner, again with the gradual displacement of Hungarians and Saxons from "higher" civil-servant positions by Romanians increasingly eager to marry women equal to their newfound socioeconomic standing. Equally important, while Saxon women were more prone to marry non-Saxons than their male counterparts, Csallner maintains that all mixed marriages were statistically less successful, and thus likelier to eventually end in divorce.[49] Interestingly though, Csallner does not discuss

48 Csallner was rather proud of these three articles, with the 1975 version of his *Meine wissenschaftlichen Arbeiten* claiming that they were "praised highly, their methodology considered exemplary," and that they had outlined the means of "strengthening the national sense of responsibility as practical measures to combat the mixed marriage by influencing internal migration patterns and job choices." Csallner, *Meine wissenschaftlichen Arbeiten* [1975], 18. For a glowing review of the trilogy, see Harasser, review of "Die Mischehen in den siebenbürgisch-sächsischen Städten und Märkten."

49 Csallner, "Mischehen Teil I," 247.

the question of how many children were born to mixed marriages or how they fared in relation to his insistence on the inheritance of talent (apart from a few general but negligible remarks at the beginning of the first installment of the series), and he made no mention of the verifiability of the racial degeneration he so frequently decried.

While this three-part article offers a truly remarkable treasure trove of statistical data, far too much to be expanded upon in depth here, but certainly invaluable to any purely demographic study, a brief survey of his individual case studies is both valuable and necessary, in that they reflect Csallner's increasing concern with the implications that ever more mixed marriages brought to bear on the Saxon nation.

In the case of Sächsisch-Regen, a quarter of the 2,214 weddings between 1895 and the first half of 1936 were with non-Saxons, while a comparison of the pre- and post-1918 numbers exhibited that fewer Saxons had married overall after the First World War, and that as a result of small families and emigration the number of marriages with Romanians had doubled, while that with Hungarians increased by 50 percent. Most alarmingly though, the number of mixed marriages had rapidly risen in the interwar period, with the percentage of Saxon men wedding foreign wives jumping from 6.62 to 17.96 percent, and that of women experienced a largely negligible increase of 1.7 percent, from 24.10 to 25.80 percent.[50]

Csallner was particularly alarmed by the results for his hometown Bistritz, although he found himself restricted to working with the data for all Saxon weddings since 1901. In this case, Csallner asserted that the emigration of young Saxon men had necessitated that Saxon women wed foreigners, and the departure of the Hungarian civil servants after 1919 led them to increasingly choose Romanian spouses.[51] Although he had hoped to find Saxon women only entering mixed marriages when despairing about the absence of Saxon suitors, Csallner was left lamenting that the intermarrying brides' average age was actually lower than that for inter-Saxon weddings (as was the case for Sächsisch-Regen). He resigned to the conclusion that "one sadly finds little of the healthy resistance to mixed marriages left in Bistritz—and not only so because of the numbers, but in what one sees and

50 Ibid., 229–30.
51 Ibid., 249.

hears when there. It is going to take a concerted educational effort to change this."[52]

A similarly dismal outlook emerged from his Schässburg case study, where Csallner had assessed all marriages between Saxons, Romanians, Hungarians, and Jews entered into between 1895 and 1936. He concluded that after the First World War the ratio of inter-German and inter-Romanian weddings had changed decisively; while the Saxons tended to marry more frequently prior to the war, the Romanians did so after—a development that underscored the disadvantageous changes to the city's ethnographic constitution.[53] Csallner argued that while fewer Saxons tended to enter into mixed marriages here than in Sächsisch-Regen, their frequency had nonetheless increased since 1919, particularly amongst the Saxon women, although the foreign spouses were usualy Hungarian. But why did Saxon women, in particular, choose to marry either Romanians or Hungarians? Csallner, as always, referred to the rapid socioeconomic rise of Romanian white-collar bureaucrats as a key factor: "The little people, these subordinates who had hardly been in respectable stations, have suddenly become 'gentlemen' [*Herrn*] with partially wide-reaching opportunities to use their newly acquired power—and they have become rich. The number of civil servants has rapidly exploded from three to eleven, and that of academics from three to twenty-six," a telltale sign of how the "Saxons have encouraged the tremendous advances made by the Romanians."[54] But why, asked Csallner, would a Saxon marry knowingly into a nation inferior to its own?[55] For one, he pointed to the Saxon brides' age being lower than average, indicating that they had less moral inhibitions. Not surprisingly to Csallner though, while all mixed marriages exhibited higher divorce rates (in this case study: inter-Saxon 2.5 percent, mixed marriages 4.7 percent), Saxon men appear to lead "happier" mixed marriages than their female counterparts.[56]

Similarly, Csallner's data for Hermannstadt for the period 1895–1937 led him to the conclusion that while the interwar period had not witnessed a

52 Ibid., 255.

53 Csallner, "Mischehen Teil II," 15.

54 Ibid., 29.

55 Ibid., 31. Why would "these German women marry non-Saxons [*Volksfremde*], whereby they not only marry out of their own nation, but simultaneously into one that, if not necessarily from a socially lower class, is in any case a nation of lower rank."

56 Csallner, "Mischehen Teil I," 35–6.

steep increase in the number of mixed marriages, other characteristic trends existed here as well: Saxon women entered into mixed marriages more frequently than men (if by a narrow margin) from the mid-1920s onwards, and they preferred Hungarian to Romanian partners prior to 1919—a trend reversed in the interwar period. In relation to the socioeconomic standing of non-Saxon spouses, Csallner found that similarly to Schässburg, the Saxon lower classes tended to intermarry more frequently—while, inversely, the growing Romanian share of the upper class did so as well. He was astonished to find that unto 1914 the average age at which both Saxon and Hungarian brides entered into mixed marriages was two years younger than that of their Romanian counterparts. While he simply refused to consider a more pronounced Romanian resistance to mixed marriages as a viable explanation for this trend, Csallner leaned heavily on his interwar data to evidence these statistics' inversion. So, suddenly, it is the Saxon women that enter mixed marriages later than either Hungarian or Romanian women, allowing Csallner to happily conclude that the age discrepancy "is largest in the distinguished German nation, where we expected to find the greatest opposition to mixed marriages anyway, and smallest in the lowly Romanian nation."[57] His analysis of Hermannstadt's marriage patterns, naturally, concluded with an evaluation of local divorce rates. Here, very much in compliance with his own views and expectations, Csallner found that although mixed marriages were more likely to end in divorce, it appeared that Saxons "got along" better with Hungarians than with Romanians.

However, confronted with the overall increase in mixed marriages, Csallner was adamant about them being the predictable consequence of the Saxon peoples' wanting knowledge of hereditary mechanisms, and of their inadequate sense of national duty. An interesting example of how this demonization of mixed marriages was brought to the public is afforded by an open letter he wrote to an unnamed Saxon woman in 1940. She had written to ask him for advice on whether to marry her Hungarian boyfriend, and despite his attempts to be polite, the message was blunt. Questioning whether she could ever be happy with a Hungarian (assuming that she wanted to maintain her Saxon identity), Csallner emphasized that as a German "girl" she had obligations to her nation that went beyond her personal

57 Csallner, "Mischehen Teil III," 257.

happiness. While she was free to marry whom she desired, the national community reserved the right to exclude her for abandoning her heritage.

> But this is not about your own personal happiness alone. You are a German girl, have blossomed out of our German nation, and are indebted to it. And it is not the individual and his/her happiness that constitute the highest of purposes, but above him/her stands the nation, and it has the right to demand sacrifices, to demand that the love for a man be sacrificed to it. This can, of course, be terribly difficult. And I do not know whether you have the strength to do so. If you do not—no, there is no need to fear that I would hate or accuse you; but you will have sinned against our nation and must endure—as you will have thus excluded yourself from the German community—that the community too considers you excluded; that is, if it is still healthy enough, and still has the necessary will to live, to do so.[58]

Csallner also ventured to query whether the Saxon woman would really want her children to be half-Hungarian by blood, if not necessarily so by culture, and he proposed that she come and live in a safe house for three months to think the matter over.

The study of mixed marriages offered Csallner the perfect arena in which to discuss the wide range of both internal and external factors he deemed so detrimental to the Saxon nation's spiritual and racial health. It illustrated his insistence on the ethnic hierarchy from Saxons down to Hungarians and then Romanians (not to mention Jews or Gypsies), and on the manner in which the latter had been catapulted into an unparalleled socioeconomic stratosphere by the state's direct discrimination of Saxon (and Hungarian) economic autarky (or, rather, exclusivity). But as his insistence on the importance of the bride's age illustrates, Csallner was primarily concerned with diagnosing the "natural health" of local moral resistance (or disdain) for mixed marriages, and with showing how that reflected upon an understanding of the dysgenic crisis unfolding around them. What Csallner failed to provide, despite his mathematical concoctions, was evidence for his central line of attack—namely, how these perceived aberrations impacted the "inherited talents" passed on to their mixed-race children.

58 Csallner, "Die Frage der Mischehe: Ein Brief," 2.

v. *Lebensraum*: Of "Foreign Invaders," Saxon Employers, and Society's Scourges, Alcohol and Tobacco

If fertility, overall, and the reproductive ratio of the supposedly inherently superior and inherently inferior, in particular, constituted the one central pin holding Csallner's ideological framework together, then the quest to rid the Saxon *Lebensraum* of "foreign" competition was its logical conclusion. While Csallner had so fundamentally disagreed with Siegmund's thesis that a growing *Raumnot* was the underlying cause of declining Saxon fertility and subsequent degeneration, the necessity of re-homogenizing and re-entrenching Saxon *Lebensraum* under threat from foreign usurpers became the basic premise for virtually all of Csallner's eugenically minded population policies. Where Siegmund demanded territorial expansion and conquest, Csallner advocated a program of ethnic re-homogenization. And if the cure lay in the cause, how did Csallner explain the gradual loss of Saxon territory?

Apart from the tremendous "external" pressures brought to bear by the Romanian state's agricultural reforms that disappropriated vast swaths of land (with the church and communal bodies bearing the brunt of this loss of income), Csallner's thought can be subsumed under three mutually amplifying "internal" themes: first, he returned to the basic presupposition that the increasing scarcity of Saxon human capital offered the ethnic others a beachhead; second, he argued that this foothold had subsequently grown as the continuing shortage of Saxon laborers led to an increasing dependence on these "invaders" to provide the labor Saxons either could not, or would not satisfy themselves. What made this situation considerably worse was that the employment of non-Saxons resulted in the automatic, unequivocal loss of national wealth, and a further loss of property. To Csallner, this amounted to a perpetual downward spiral, where "foreigners" used the capital they had initially taken out of the Saxon economy to further degrade its integrity and homogeneity (thereby, amongst others, opening the door to mixed marriages). In addition to Csallner's omnipresent calls for a higher birthrate and to his disdain of "ethnic others," a third theme emerges from his discourse, one inspired by Siegmund's crusade against the consumption of alcohol and tobacco, which robbed the national purse of dearly needed funds, beyond the damage inflicted on its collective health.

Csallner dedicated comparatively little of his own time, and especially so in relation to his "sociobiological studies" on differential fertility, to the in-depth study of the extent to which the Saxon *Lebensraum* had been eroded over time. He approached the theme from two distinct avenues: engaging in predictions of how patterns of encroachment could reveal, on the one hand, future trends, and local micro-level patterns in property acquisitions and sales, on the other.

A particularly useful example of the first approach is offered by Csallner's interpretation of two of the LSS's most intriguing maps, dating from the late 1930s. The first one, *Transylvania's Germans* (*Das Deutschtum in Siebenbürgen*), provided a visual representation of the German numeric strength in the region, or more precisely, the relative degree of multiethnicity. While this map regrettably does not provide as detailed a historic overview as Siegmund's *Deutschen-Dämmerung in Siebenbürgen* does with regards to the growth or decline of specific ethnicities, Csallner nonetheless deduced that: "On the Siebenbürgen map we can see that while all of the villages around Hermannstadt, with the exception of Hammersdorf, still have German majorities, they are in danger of being cut off from the further eastern and northeastern area with German majorities, in most settlements by a bar—of once mainly or purely German villages, where they [the Germans] are now a minority and threatened to be driven further back still—stretching from the northeast to the southeast."[59]

While Csallner was fearful of an eventual splintering of the Saxon *Lebensraum* more widely, he was adamant that Hermannstadt found itself increasingly besieged by a belt of entirely non-Saxon villages (which were not included in the Siebenbürgen map). LSS's second map was hence particularly alarming, as it illustrated how a mere handful of settlements boasting a Saxon majority were effectively "defending" Hermannstadt's southern and southwestern "frontier" from ever-further-encroaching Romanian colonies. With the rather sensationalist-sounding title *The Impending Isolation of Hermannstadt's German Areas by Advancing Romanians* (*Drohende Abschnürung des Hermannstädter deutschen Gebietes durch vordringendes Rumänentum*), Csallner wanted to underline the paramount importance of internal colonization schemes geared towards strengthening such Saxon bastions—before

59 Csallner, "Das Landesamt," 8–9.

they succumbed to the "foreign flood" that had already begun to consume Hermannstadt. In Csallner's own words, this map offered a rather clearer representation of the looming danger, "because it also includes the towns with entirely foreign populations. This determines the necessity to use all available means to not only prevent this strip of settlements with a German minority from weakening any further, but to do everything possible to create a German majority again."[60]

One of the very first tasks Csallner tackled through his nascent Race Office in 1932 was the compilation of village maps to serve as the groundwork upon which around two hundred studies of local property markets were to build.[61] Csallner was interested in collating data on a wide range of issues, ranging from the rather straightforward question of total property and which ratio thereof was owned by Saxons; to an account of recent transfers and who had "profited" and who "lost"; the impact of the state's agricultural reforms on plot size and ownership; to what end each property was being put; and even the rather difficult to qualify question of whether properties should be deemed "safe" or "endangered."

In 1934, Csallner's workforce was supplemented by the Subject Service of the Combined Students in Romania (Fachdienst zusammengeschlossener deutschen Hochschüler Rumäniens), which offered their help, and whose students he swiftly dispatched to about fifty communities to evaluate the health of their respective property markets.[62] Despite being embroiled in two lawsuits, Csallner was convinced that this tedious but necessary work bore its richest fruit not merely in terms of the actual, physical research conducted, but due to its educational function of rearing the nation's future leaders and imbibing in them an intrinsic understanding of the battle for survival at hand: "The greatest benefit, though, is that the students, when they arrive in our communities and [...] have to work their way through them house by house, look deep into our nation's life, into her wants and battles, learn of all the various promising ways to work for our nation [*Volkstumsarbeit*], and thereby not only learn about the community being studied, but learn to see and think national-biologically, like future leaders in the

60 Ibid., 9.

61 Csallner, *Meine wissenschaftlichen Arbeiten* [1944], 11. See also Csallner, *Meine wissenschaftlichen Arbeiten* [1975], 12; and Csallner, *Zur Verleihung des Kulturpreises*, 5.

62 Csallner, *Die volksbiologische Forschung*, 103–104.

fight for our continued existence have to see and think in order to be able to lead properly."[63]

In addition to the educational value these projects entailed for aspiring researchers and leaders, Csallner was, as always, exceptionally eager to engage with local priests and teachers for similar reasons. While these were largely tasked with collecting data in rural areas, Csallner was also to rely heavily upon the *Nachbarschaften* to do the same in towns and cities (although almost exclusively in Siebenbürgen). Eager to increase the number of completed local surveys, Csallner led two training sessions on how to carry them out, one in Klausenburg and one in Bucharest, bringing the total number of coworkers up to a rather impressive 120.[64] Much to Csallner's delight, a number of related studies were soon published independently of his own work, amongst which we find, for example, Hans Kaufmes's study of property transfers—if less detailed—for most of the Burzenland.[65] Father Gustav Arz, who was to head the LSS's Department for Parish Registers, published on "Protestant Church and Germanness in Hermannstadt" ("Evangelische Kirche und Deutschtum in Hermannstadt") in 1936, and further works were advanced by Csallner's student Franz Wiegel, and by the judge Dr. Fritz Mild, who conducted a survey of Kronstadt.

Concurrently, Csallner tackled what seemed to be the ever-growing problem of rural employment patterns.[66] While he never completed his work quantifying the damage inflicted by the employment of non-Saxons as agricultural laborers, maids, and so on, they appear as the rural equivalent of his equally vocal concern that Saxon handymen and traders in towns were training far too many "foreign" apprentices. Csallner insisted that this situation was the foreseeable result of insufficient internal colonization, on top of the disastrous decline in rural fertility, but he also attributed it to widespread indifference and a misplaced sense of pride. A wonderful example of Csallner's efforts to encourage Saxons towards labor-market autarky is provided by his friend Wilhelm Klein's account of the lukewarm reception Csallner's speech received from a crowd of peasants in Alzen, whom he implored "to abandon the damned prejudice that poorer peasants thought

63 Ibid., 104.

64 Wolff, *Ein Jahr Volksgemeinschaft*, 59.

65 Interestingly enough, Kaufmes was the director of the Ackerschule in Marienburg, and later the Volksgruppe's national peasant leader (*Landesbauernführer*).

66 See Csallner, "Die volksbiologische Lage," 10–1.

it beneath their dignity to send their sons and daughters to help—for pay, of course—the wealthier peasants with their harvests once they themselves were done, rather than force them to rely on gypsies. The excellent speech fell upon deaf ears. The Saxon peasant, it seemed, held on to his prejudices, and much more tightly so than to tradition."[67]

Using Holzmengen as his case study, Csallner's 1935 poster detailing the settlement's employ of "foreign servants" (*volksfremde Dienstboten*) illustrates this point exceptionally well. Subsuming three intrinsically related themes, Csallner set out to quantify precisely how many non-Saxons worked as servants; to calculate the extortionate amount of capital lost to ethnic others; how much land it could have secured rather than endangered; and how much money could have been saved had all the Saxon youths worked. In effect, Csallner concludes that if the "spare" eleven male and eleven female youths he deemed "available" to work had actually done so, the local Saxon economy would have "lost" 40,000 rather than the staggering 183,000 lei in salaries for foreign labor—with which they could have bought either twenty-six joch worth of fields (where one joch is the equivalent of 0.5755 hectares) or even two of the 195 local farms. Returning to the question of internal colonization as a defensive policy to fill nationwide gaps in the labor market, Csallner's poster concludes with the thought that, in fact, seventeen girls could have gone to work as maids in Hermannstadt, thereby not only dispensing with the neeed for non-Saxon staff but actually earning a significant portion of their dowries themselves.

But there is also a wider theme at work here, namely Csallner's deep-seated disdain of what he considered to be the degenerative influence suffered by young Saxons working abroad—in particular, by those who had emigrated to the Romanian Old Kingdom. His interest in this aspect of the labor market was not merely informed by economic concerns, but by the need to tabulate the availability of "spare" or "excess" human capital that could be diverted to other Saxon towns rather than be spoiled by prolonged exposure to Romanians above all else:

> First, we have to see where we need people and where we can get these people from—and that is what my surveys wanted to contribute. Simultaneously, I also wanted to find out what communities do the Saxon girls, especially,

67 Klein, ". . . lies Kirchenteppiche pfänden."

> leave to serve—sadly Bucharest or other places in the Old Kingdom instead of Saxon cities—where they are easily corrupted [*verderben*], frequently even become ill, and when they return do not bring a good spirit [*Geist*] with them; so that we may know how prolific this evil [*Übel*] has already become, and where we especially need to battle it.[68]

In tandem with Siegmund, Csallner had become increasingly disturbed by the Saxon consumption of alcohol and tobacco, both of which were deemed to be amongst the most prolific and deeply engrained of all socially degenerative forces. A rather obvious example of this is afforded by another of the LSS's posters, Dr. Josef Bacon's analysis of Schässburg's 1921 expenditures on alcohol in relation to their investment in church and schools. Calculating in kronen, Bacon's bar graph reflects the immense disproportion between the money that Saxons had spent on alcohol (wine and sparkling wine totaling 2,812,800 lei, beer 494,364 lei, spirits 1,052,270 lei) and tobacco (3,000,000 lei), which totaled at 7,359,434 lei, and a meagre 907,000 lei paid to the church (church and school taxes amounting to 800,000 lei, and school fees to 107,000

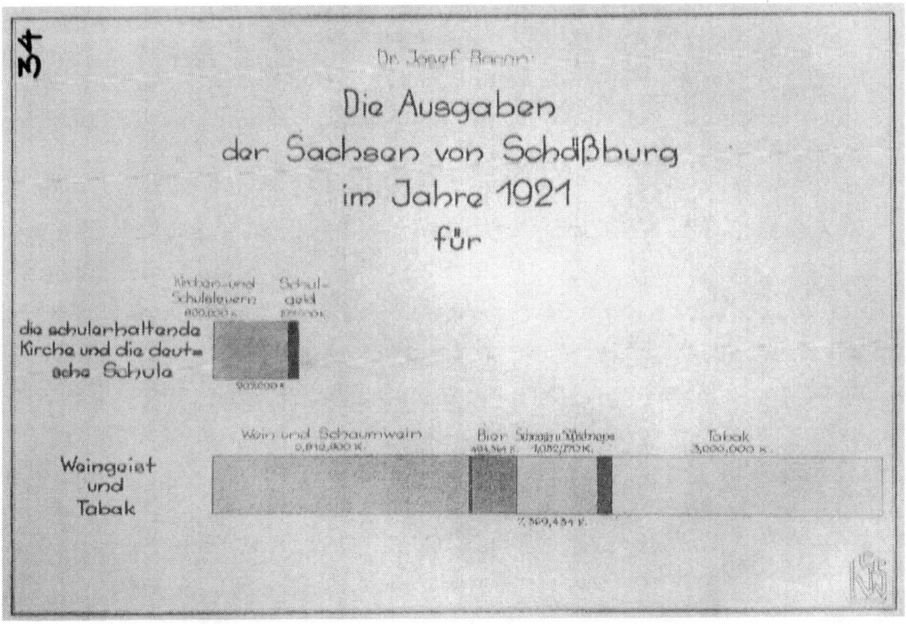

Figure 5. Dr. Josef Bacon: The Schässburg Saxons' Expenditures in 1921.

68 Csallner, *Die volksbiologische Forschung*, 103.

lei). Schässburg's Saxons had spent about eight times as much on agents of degeneration as on the existentially vital confessional schools.

Csallner's understanding of the immense damage inflicted by alcohol and tobacco, and the necessary battle against them, also assumed a distinctly political and ideological dimension. In a truly remarkable and unusually aggressive in tone article published in the *Selbsthilfe* in 1932 under the title "An Open Letter to All Who Want to Be National Leaders," Csallner reflects on the significant benefits that a drastic reduction in the consumption of these societal ills could yield for a Saxon national economy straining to subsist under the Romanian "yoke"—one, indeed, not merely neglected but driven to the brink of collapse by a premeditated and active state agenda:

> We, as a nation, are in serious danger of collapsing, and perhaps very soon. The most apparent threat is an economic collapse. But this would also entail our cultural decline, and maybe even destruction, yes indeed: it would mean our destruction as a nation. But the economic crisis has only become so substantial and dangerous because we have failed morally [*sittlich*], and continue to fail. In addition, though, we have a particular crisis—namely, that we have been seriously robbed since we belong to Romania, are incessantly sucked dry for foreign purposes, and that the state refuses us the bulk of what it owes us from our taxes. But the more we talk and write that we will be ruined, all the more tenaciously will the state continue to deny us our rights—because we are supposed to vanish![69]

It is within this framework of economic degradation that alcohol and tobacco become particularly destructive influences. Csallner unleashed a torrent of abuse directed at both the church and the political establishment for having failed to take the lead in the fight against these economic and spiritual "plagues": "Only grave mental bluntness [*Stumpfheit*] and moral crippling can today not yet deem it absolutely necessary to smoke and drink as little as possible—if it is impossible to stop altogether—for the good of our nation and its future, due to economic as well as moral, national, and hereditary reasons."[70] It is only logical that Csallner concluded his article with an avid call on Saxon priests and politicians alike to at least publicly denounce

69 Csallner, "Offener Brief," 2.
70 Ibid.

alcohol and tobacco as the agents of degeneration they were, if not join the International Order of Good Templars (IOGT) that he himself had joined through one of its highest ranking members, Siegmund.

Csallner had a peculiar relationship with the IOGT, and while he published some of his most interesting pieces in its organ, *Der Aufbau*, his "The Saxon Priest as a Good Templar" is a characteristic example of the hostility he complained its calls for abstinence evoked amongst the wider populous.[71] Perhaps most importantly at this point, though, was Csallner's insistence on the priest's role-model function in Saxon society, and on the clergy's particular duty to battle these substances' abrasive effects. He called upon the clergy to not only mend its own ways, but join the order as a public act of commitment to the cause. This was all the more essential as the priest remained a figure of authority in a time of societal turmoil: "Because our nation, no matter how stormy and inflamed it may be, remains full of honest respect for its priests, and is still so used to believing what we tell and give it, that no one would have an easier task convincing it to give up drinking as the Saxon priest would."[72] It were hence the priesthood's mission, should it want to remain the nation's leader, to be the vanguard in this battle, as Csallner himself had joined the order precisely because he was a priest.[73] The church was the default agent via which Csallner aspired to infuse his definition of the cancer and regenerative cures into the Saxon national consciousness. What is more, he sought to employ the church's vast structural network as the key legislative and executive engine of his eugenic population policies.

71 "Dare I become a Good Templar, then it can well happen that I find myself at war with the family, and from all sides the good friends will storm: Man, that is the last thing you needed—you made a fool of yourself—you have only damaged yourself!" Csallner, "Der sächsische Pfarrer," 4.

72 Ibid., 6.

73 Ibid., 7.

CHAPTER III

ALFRED CSALLNER IN SEARCH OF EUGENIC SOLUTIONS AND INSTITUTIONAL MEANS

What measures did Alfred Csallner propose to counter this unfolding dysgenic crisis, by what means did he hope to translate them into reality, how were they going to be executed, and, critically, by whom? Apart from his efforts to enlighten and awaken the masses to their national race-hygienic duties through lectures, workshops, work groups, and other private initiatives, Csallner accorded the church a significant role in managing the supply and demand of various urban and rural professions, in the manipulation of internal migrations to strengthen endangered settlements, and in imposing punitive measures. It was only logical that Csallner, like Siegmund, embarked on a mission to woe and reform the church from within during the 1920s, before turning to the Self-Help as a vehicle for realizing his agenda in the 1930s—thereby completing the transition from clerical to political means following the remarkable period of overlap when Siegmund, Csallner, and Fabritius collaborated in the church's Welfare Committee. This is by no means to say that Csallner was himself not genuinely devout, or utterly convinced of the compatibility of religious faith with racial hygiene. On the contrary, Csallner wanted to reinvent the spiritual and educational remit of priesthood itself, while employing the church's hierarchy towards translating his eugenic ambitions into legislated reality. In other words, Csallner understood the Saxon church as a *Staatsersatz*, an organizational superstructure that needed to appreciate that eugenic imperatives overrode religious dogma, and whose priests and teachers needed to embrace eugenic tenets if they were to fulfil their duty and salvage their individual flocks from perpetual degeneration.

1. Eugenic Missionaries: Visions of Priests Old and New

Csallner's ideal priest was hence more of an academic than a theologian, and he wrote prolifically about the pitfalls of the increasing number of priests trained in seminaries rather than a wider range of academic disciplines (notably, philosophy). Csallner harbored a deep-seated disapproval of the ongoing proliferation of "seminarists," whom he considered by definition inferior intellectually and therefore hardly equipped to lead their flocks during a period of existential crisis. Indeed, as his brief summary of the uncompleted "sociobiological" study of the priestly profession's reproductive trends illustrated, Csallner insisted that "seminarists" were frequently the offspring of families of lesser hereditary value than their academic counterparts, and that their spread consequently implied the entire caste's creeping loss of racial worth. Csallner maintained there was a particular need, first, to drastically increase the number of academic priests; second, to increase the overall number of teachers and priests and ensure that most villages had at least one of each rather than a hybrid role; third, that the priests in particular should awaken to their new core mission as "torchbearers" of the eugenic truths that alone could avert an increasingly desperate dysgenic crisis; and, fourth, that they lead the struggle towards regeneration by example, by having more children themselves, abstaining from drink and tobacco, engaging with Csallner's local "work groups," and otherwise stand at the forefront of "national-biological" and "sociobiological" research.

A particularly insightful example of the extent to which Csallner hoped to redefine the duties, not to say the very nature of the Saxon clergy, is afforded by his essay "Torchbearers," initially published by the church's official organ, the *Kirchliche Blätter*, in 1928.[1] In 1926 and 1927, the Landeskonsistorium had been persuaded by its Welfare Committee—i.e., by Siegmund and Csallner—to send three race-hygienic texts to its local parishes for discussion at its teachers' and priests' general assemblies.[2] In and by themselves interesting choices, given that they exemplified a rather more right-wing approach to eugenics compared, for example, to Wilhelm Schallmayer, the three books were Andreas Thomsen's *Nations' Elapsing and Emerging* (*Der Völker Verge-*

1 Csallner, "Fackelträger."

2 LK directive ZL. 2898/927, dated June 17, 1927.

hen und Werden); Hermann Werner Siemens's *Foundations of Hereditary Theory, Racial Hygiene, and Population Policy* (*Grundzüge der Vererbungslehre, der Rassenhygiene und der Bevölkerungspolitik*); and Fritz Lenz's *The Biological Foundations of Education* (*Die biologischen Grundlagen der Erziehung*).

The reports subsequently submitted offer an intriguing assortment of views, ranging from an enthusiastic embrace, at least publicly, of race-hygienic ideas to outright hostility—a snapshot of the perceptions of local priests and educators on the eugenic mantra of encouraging the fit.[3] They mostly agree on a clear need for both professions to engage with race-hygienic ideas more broadly, but also consistently voice serious concerns about the pitfalls and potential implications of such an unequivocal insistence on the inherited talent. This set of reports offer a tantalizing dimension to the hallmark nature-versus-nurture debate of interwar eugenics, and yielded a variety of calls for a broader, quantitative population-growth policy, and the discouragement of mixed marriages, amongst others. Characteristically and unavoidably, such practical measures to boost birthrates would invariably rest with the church, whose Christian morality could either be invoked as a call to arms or the bedrock of principled refutation.

So, for example, we find a ringing endorsement of Siemens's views on the "struggle for life as a battle of births," and another full of criticism of interwar individualism that for its part concludes: "Pushing back the personality cult [*Personenkult*]—which is seriously degenerated in its current form, in contrast to the duties to future generations—can only be effectively achieved through man's religious disposition. That is why the doctor and researcher, together with the priest, must demand the same thing: more religiosity."[4]

In contrast, the report submitted by Dr. Alfred Roth argued that the need for race-hygienic measures amongst the Saxons, albeit instructive and important, was not as urgent as in Germany, because "our peasantry, as a breeding factor [*Zuchtfaktor*], is not yet [sunken] as low as the German proletariat."[5] Nonetheless, Roth continued, the church could realize Siemens's call to support large families, "in particular the culturally valuable classes," through its salary schemes and church tax system. This preference

3 While most are held in archival sources, a few were published, for example Martini, "Pfarrversammlung im Unterwald"; and "Aus den Verhandlungen der Bukarester Pfarrkonferenz."

4 NA Sibiu, CGR.K.23, p. 447.

5 Ibid., 440.

for the professional classes is also evident in the Schässburg report written by Friedrich Czikeli, which demanded more concrete measures from the church, so to encourage early marriage and large families through its salary and child-benefit schemes (the latter was limited to three children), by giving large academic families preferential treatment when assigning accommodation, and by amending the school curriculum and teaching children to live frugally and simply amongst others.[6]

While the numerous local reports offering their views on these three books' compatibility with church teaching and on their classroom suitability revealed remarkably varied responses, Csallner's "Torchbearers" was his reply to some of the more apparent criticisms voiced by the clergy against the moral and practical implications of race-hygienic measures, and offers his most outspoken attempt to convince them of their merits, his bid to transform the Saxon clergy into "torchbearers" of this "new" ideology as proclaimed by its "prophets": "The preachers of these new truths—no, of these old truths that we have only now begun to understand—they are prophets called upon to refashion the way the future will feel and act."[7] Rejecting the thought that race-hygienic "truths" were in themselves "terrible," Csallner proceeded to argue that they constituted the only real measure by which the Saxon nation could arrest its racial degeneration, and could even turn it around, eventually, into a racial regeneration. Infused with religious piety, the following is a surprisingly concise summary of his approach to racial hygiene, its merits in pursuit of racial salvation, and its compatibility with Christian morality—indeed, its divine calling:

> The only terrible thing is that we have, until now, unaware of these laws and in complete ignorance, slipped onto a sloping road. One that has resulted in such an accelerated process of racial degeneration, that even with the utmost of vigor it will take a long time to truly bring to a standstill before it can be turned around into regeneration. [. . .]
> And it is terrible that so many still turn a blind eye to these truths, and endeavor to keep others from recognizing them. That by severely complicating—maybe even retarding—our recovery and salvation, they might perhaps make it entirely impossible. [. . .]

6 NA Sibiu, CGR.D.XIV.37, p. 37.

7 Csallner, "Fackelträger," 2.

> But the laws of heredity are anything but terrible. They are holy and great, so much so that we piously shudder in front of them, as if looking into the starry sky: Lord, how grand are your achievements, so great and many![8]

That said, it is of course little surprising that Csallner culminates in emotive calls to the clergy to live up to their particular role in Saxon society as its educators and spiritual shepherds, by taking up the race-hygienic cause: "But we need torchbearers on this road forward and upward. And we, as priests, are all the more called upon to be our nation's torchbearers."[9]

Csallner then proceeded to qualify what to him had always been the engine driving racial degeneration—the Saxon nation's continuous loss of hereditary worth. Drawing extensively from Heinrich Wichern's *Sexuality and Population Policy* and its defense of "breeding" practices as preventative measures against "inferior" individuals,[10] Csallner furnishes the reader with a very concise but accurate summary of the problem perpetuating Saxon degeneration, while postulating its race-hygienic solution:

> The most important concern, though, is not with curtailing the inferiors' reproduction, but the protection of the currently rapid extinction of the highly- and best-endowed amongst us. They are the bearers of progress and irreplaceable. If a nation loses them, it is itself lost. That is why we must nurture family values, reawaken the joy and willingness to have more children—particularly amongst the better and best racially. And especially these must realize that they, more so than all others, have a duty onto God to maintain and multiply their superior hereditary traits.[11]

Not only does their societal standing demand that the clergy embark on a holy mission as torchbearers of a race-hygienic discourse perfectly in tune with Christian morality, they must also be its arsonists in their congregations, by encouraging those deemed racially valuable to recognize their importance to the nation's very survival.

8 Ibid., 3.

9 Ibid., 7.

10 Wichern, *Sexualität und Bevölkerungspolitik.*

11 Csallner, "Fackelträger," 6.

II. Csallner's Population Policy Proposals and the Church

Part of the realities of the 1920s church reforms was the need to drastically cut costs where possible. Having lost a substantial amount of its property portfolio and fiscal power to the state's agricultural reforms, the Saxon church saw itself increasingly pressed to restrict its expenditures. Nonetheless, Csallner's publications are peppered with criticism of these reforms, and littered with his own reform proposals (in most cases fully costed at that). He shared a widely held view that the church's wage and welfare structures ought to be revised in light of its redefined *Staatsersatz* mission, and in order to subsidize his various eugenic agendas—from rewarding large families (or, rather, penalizing small ones) to employing more academic priests and teachers.

In 1928, Csallner called on the church to provide the funds necessary to employ both a priest and a teacher in every Saxon settlement of a certain (but unspecified) size. The brief, seven-page-long article "On the Minimum Requirement of Priests and Teachers and a New Solution to the Wage and Distribution Question," published in the *Kirchliche Blätter*, presupposes that "despite all exceptions on either side, no one can possibly deny that academic priests are better than seminary ones," and that "whoever has understood the circumstances knows this anyway."[12] Csallner insisted that the new salaries created by this two-percent increase in staff would barely be felt by the church's purse, and that regardless of how expensive this new "armor" would be, the realities of their embattled position rendered any other solution both "irresponsible and disastrous."[13]

His central concern hence always lay with these torchbearers' national duty to further and encourage their flock's willingness and inner desire for more children. All of his subsequent calls on the church, for example to reduce the taxes and tuition fees asked of large families, were not merely of fiscal value but a "moral support." The church had to preach the "blessing" inherent to rearing at least four children, and this unequivocal truth needed to be instilled in their young, preferably in the confessional kindergartens, but in schools at the very latest.[14] In effect, the goal was "to fill the youths

12 Csallner, "Vom Mindestbedarf an Pfarrern," 2.

13 Ibid.

14 Csallner, "Förderung des Kinderreichtums," 175.

with the happy will to have more children, and to strengthen the girls, in particular, so that once married they are strong enough to resist when their mother, mother in law, good friends, or even their husband push for family planning or urge them to have abortions, or any other attempts to give them good advice."[15]

Woven through all of his proposals were calls for higher child-benefit payments in tandem with the revision of the church's tax regime in favor of large families. In his poignant 1927 article "Children as Luxuries,"[16] lamenting the existing system that limited payments to a maximum third child, Csallner repeatedly warned the church of the disastrous consequences their reduction of child benefits would entail, and did so in vain. The following year, he published a scathing critique of the new system, which he was convinced would further disenfranchise large families.[17] Advancing three main arguments, Csallner claimed it would, first, lead to a further reduction in birthrates all the more detrimental to the cause as, second, teachers and priests needed to lead by example. To make his third point, Csallner returned to his theses on the concentration of particularly valuable hereditary traits in selected professions, arguing that "due to societal selection the individuals who rise to become priests and teachers generally have above-average talents, and the children of priests and teachers also tend to be talented above average."[18] Csallner chastised the church for what must necessarily result in the "progressive loss of irreplaceable, best hereditary traits, an ongoing degeneration [*Entedlung*] of the entire nation, and its ongoing racial decline."[19]

Undeterred, he advanced his own reform package, along with a surprisingly detailed costing, one he introduced with: "I demand: married employees are to be paid more than the unmarried ones. The child benefits may not be limited to a certain number of children but must be calculated in such a way that they increase with every child, and hence fulfil their purpose better the more children an employee has."[20] Having compiled data on the age and family size of 1,084 church employees and their total of 589 children from church archives, Csallner based his subsequent calculations on vari-

15 Ibid., 176.
16 Csallner, "Luxuskinder."
17 Csallner, "Vorschläge zur Neuregelung."
18 Ibid., 2.
19 Ibid., 3.
20 Ibid., 5.

able "units" of tax and "units" of benefit payments. Setting the unit level of this tax (to be deducted from their pay by the church) to increase with age in the case of unmarried men, while staying at a flat rate for unmarried women, and halved for married but childless employees, he proposed to increase the level of benefit by one "unit" for each child. What is more, and although even Csallner realized the impossibility of doing so, he ventured to propose that the level of benefit could be raised considerably higher, and achieve considerably more substantive results, if it was set to begin with the third (rather than the first or second) child.[21]

That same year, Csallner's Society of Child Enthusiasts also submitted a petition to the church proposing several substantial reforms to its payroll system, the centerpiece of which was an "equality fund" (*Ausgleichskasse*).[22] With the rather unwieldy title of "Draft law on the creation of fiscal equality between employees with few or no children and those with many within the Protestant national church in Romania, towards the protection and furthering of large families," the fund had recast the carrot as a stick—one designed to tax all church employees out of any fiscal advantage that having less than four children afforded.[23] To substantiate the need for such a drastic revision of church policy, the *Denkschrift* bristled with the usual and customary predictions that the Saxon nation's status as a *Kulturvolk* was placed in peril by the rampant proliferation of a family-planning plague that had in itself produced an abundance of mentally deficient single children, and that in the absence of an independent state they must resort to the Saxons "being so wonderfully organized, and still too easy to influence and lead."[24] To that end, Csallner's society demanded the LK grasp the opportunity afforded by its revision of church pay packages to advance the necessary cause of drastically increasing fertility rates and the creation of a new financial equilibrium between large and small families:

> Our entire nation cannot possibly not care when one of its branches—one whose inner worth, innate and inherited, and especially rich in above average spiritual and moral attributes—is condemned to wilt, and thereby renders the entire nation poorer of high quality hereditary traits, and decays

21 Ibid., 9.
22 Csallner et al., *Denkschrift*.
23 Ibid., 8.
24 Ibid., 4.

> racially, and consequently culturally, too; or [not care] whether she blossoms and grows and thus helps to further ennoble [*veredeln*] our nation racially. And our nation cannot possibly not care if precisely those of its leaders to which it looks most are forced to give a bad example of national suicide [*völkischen Selbstmordes*]—or whether they will be able to be a good example of strong growth and future racial improvement [*Aufartung*].[25]

With the priest (preferably an academic one) stylized as the nation's most valuable leader and role model, the church's population policy needed to strive towards creating a climate in which parents could not only feed and clothe larger families, but do so to the same standard as they would with only one or two offspring. The proposal subsequently called on the church to feed its "equality fund" by withholding a staggering 30 percent of the wage of single, divorced, or widowed employees; 20 percent of that of married employees without children; and 10 percent of that of married employees with only one child. According to Csallner, this system would furnish a powerful incentive to found larger families, as the benefits would only be paid to families with at least three children. Very much to Csallner's consternation, the church did not approve of his proposals, and he began looking elsewhere for support.

The following year, in 1929, the VdK collectively joined Fritz Fabritius's Self-Help movement, and Csallner forwarded another proposal, this time calling for a revision of the church's electoral system. While neither detailed, nor elaborate, the two-faceted justification Csallner offered in favor of the introduction of a family-based voting system where parents could vote on their children's behalf is instructive of the politicization and ongoing radicalization that his eugenic discourse was undergoing at this time (all the more so as Schunn later adopted it into the movement's political program). While Csallner espoused the various symbolic and social virtues of tipping political power in the favor of large families, he was equally indignant that "it is obvious that the introduction of such an innovation, no matter how beneficial, could only be achieved here or elsewhere—unless it was simply ordered by a dictator—after a bitter struggle, one that would be

25 Ibid., 6.

the harder the longer one waits."[26] Apart from the difficulties of introducing and managing such a system in the absence of an authoritarian political regime, Csallner considered a second possible objection to his plan: because poorer, and thus hereditarily inferior families tended to be larger than the wealthy, and thus hereditarily desirable ones, such a voting system would create a political imbalance in favor of the least valuable portions of society. And while Csallner conceded that could happen, he insisted that "we must accept and live with this threat"—albeit one somewhat peculiarly mitigated with the assertion that talent was distributed more widely amongst the Saxons than in Germany.[27]

Three days after Csallner had enthusiastically announced the creation of his Self-Help Race Office in the pages of the *Selbsthilfe* on October 1, 1932, he submitted one of his most radical eugenic policy proposals to the LK's Settlement Committee (Siedlungsausschuss) for consideration.[28] Remarkably similar in nature and scope to Siegmund's Bodenschutzverein, Csallner called for the creation of "hereditary leases" (*Erbpachtstellen*) owned and managed by the local church authorities under the surveillance of a new central Settlement Office (Siedlungsamt). If only three pages long, this is a truly remarkable letter, in that Csallner's proposal was designed to address and combat virtually all of the causes and consequences he ascribed to the cataclysmic degenerative crisis beleaguering the Saxon nation. Despite the length of his justification for the unequivocal and immediate need for the implementation of his "hereditarily lease" program, its language alone merits a detailed consideration:

> I present here, in rough strokes, a settlement plan the realization of which—to mention only the most important aspects—would help to
> 1. Protect endangered *Lebensraum* and regain that already lost,
> 2. Combat declining fertility and encourage an increased reproduction amongst our nation's racially most valuable parts, and to thereby at least stem our nation's ongoing degeneration [*Entartung*], if not turn it around into a regeneration [*Aufartung*],
> 3. Use our labor better than we have to date,

26 Csallner, "Geburteneinsturz und Familienstimmrecht," 5.

27 Ibid.

28 *Csallner an den Siedlungsausschuss*, October 4, 1932, NA Sibiu, Volksgemeinschaft, D.XIV.850, p. 40.

4. Increasingly render the many foreign laborers that we currently employ superfluous, and thereby not only offer our national brothers the sources of income and job advancement we are currently giving to our enemies in the struggle for survival, but to dam the currently incessant flood of foreign nations into our settlement area—yes, perhaps even bring it to a standstill, maybe even reverse the flood [*Zurückfluten*],
5. Combat the plagues of alcohol and tobacco, and to thereby preserve all the national wealth and health and moral strength that we are currently wasting year upon year, and to place them in the service of our nation's growth and regeneration [*Ertüchtigung*],
6. Stop the de-Saxonization [*Entsachsung*] of our communities and cities and transform it into a de-foreignization and re-Saxonification [*Wiederversaschsung*],
7. Allow our nation to become strong and healthy again.[29]

In other words, Csallner believed the introduction of such "hereditary leases"—the plots of which should not be "too small," as he believed a lower workload would foster laziness—would help combat and even reverse the "flood" of ethnic others, undermining them economically, racially, and territorially; would preserve the fiscal losses currently incurred at the hands of the "curse" of alcohol and tobacco consumption; and most importantly, would bring about a racial and spiritual regeneration.

Csallner did not propose any clear mechanism through which to fund these purchases, but he was explicit about whom they should be leased to, and under what conditions. Presupposing a vigorous vetting process, prospective settlers had to be racially well endowed and healthy physically, sound in spirit and character. The settlers had to either already have four living children, or come from large families, with the promise of becoming so themselves after a maximum of ten years of marriage. Applicants had to have worked as a servant, maid, or agricultural laborer for an unspecified minimum time, and were prohibited from smoking or drinking.[30] An added incentive to have as many children as possible was built into the lease payment scheme, with the rent decreasing in relation to the number of children the family had. Furthermore, and explaining why he proposed that the

29 Ibid., 1.
30 Ibid.

church oversee this scheme, Csallner also stipulated that the parcels had to be inherited intact to one child alone, who must in turn fulfil the same requirements demanded of the parents.[31]

Considering the rather radical and explicitly race-hygienic nature of this proposal, it does appear astonishing that, according to Csallner, "the Settlement Committee had in principle agreed with my proposal, which it declared to be executable."[32] In 1934 though, when the prospect of buying a plot previously part of a Hungarian noble's estate in Billak by Bistritz offered itself, and the necessary funding appeared secured, the deal eventually failed to materialize. Regardless, Csallner was not one to be disheartened, and so he simply resubmitted his plan for "hereditary leases" in 1938, with equally little success.[33]

Although confident that such measures could benefit many an embattled Saxon settlement by bolstering their vicinity with "eugenic fortresses," Csallner was certain that many others were already "lost." He went as far as to propose the evacuation of Saxons from areas in which such schemes would bring no tangible improvement, and to deploy them in an effort to plug the gaps in an increasingly patchy *Lebensraum*:

> It is urgently necessary to resettle large portions of our ethnic community. The Germans in Bessarabia, Dobrudja, the Buchenland, as well as in the Romanian "Old Kingdom" are endangered—by factors that I do not want to discuss here—and so seriously threatened that they will hardly be able to survive by themselves [. . . There are also 'lost bastions'] where it makes no sense to artificially sustain them by investing money, work, and people into them, and with whom we could permanently strengthen still-salvageable places ,or create strong ties between the individual pieces of our so terribly shredded settlement area. I would very much like to pursue these questions and [. . .] work on finding solutions.[34]

Naturally eager to lead the statistical surveys necessary to prepare such a massive undertaking (as indeed he would, in November 1940), any policy involving such a substantive internal colonization scheme was, in effect,

31 Ibid.

32 Csallner, "Das Landesamt," 21.

33 Ibid., 20–1.

34 Ibid., 21.

impossible to carry out. Nonetheless, it aptly illustrates the lengths to which Csallner was willing to go in his bid to re-homogenize and purify what could be salvaged.

Probably the most important of Csallner's envisaged projects came in the form of a "hereditary fitness card index" or "archive" (*Erbtüchtigkeits-kartei/Archiv*) he hoped to set up in collaboration with the church and the medical association.[35] Csallner wanted to tabulate the hereditary worth and talents of specific families, in an attempt to influence individual career choices and to encourage higher birthrates where desirable. Insisting on the low cost and comparative ease with which the necessary data could be extracted and compiled from the existing family databases held by the church,[36] Csallner was adamant such a hereditary archive was something of a minimum requirement compared to the vast leaps forward made by the Third Reich. And while "naturally, we do not have many of the helpful things we would if we lived in the German Reich: hereditary courts, jails, hospitals, medical insurance, and so forth," the local priests and teachers alike must nonetheless recognize the utility of such an archive created and managed by the church.[37] As for the archive's remit, Csallner wanted parish priests to collate data on any given individual's school grades, medical evaluation—more precisely, the hereditary or *erbärztliche* assessment, preferably by a German doctor—and the level of land ownership (with regards to peasants).[38] This information would then be transcribed unto forms grouped by family (with a new file created when a child got married), and updated annually. Structurally, Csallner wanted to spread this "hereditary archive" over ca. 250 local parish offices that could, when needed, quickly and efficiently compile any given individual's/family's ancestral or hereditary tree. Clearly, such an archive would have served to substantiate calls for a considerably more invasive approach to internal migrations, if nothing else, in an active effort to correlate the pool of available skill and talent to immediate shortfalls in the job market, and hence battle the Saxon economy's need for "foreign" labor.

35 An idea he had already expounded in Csallner, "Familienbuch—Erbarchiv."

36 Ibid., 1: "And they [priests and teachers] will be all the more willing to help and look forward to helping, [...] as we can create our future hereditary archive, reliable and comprehensive, more cheaply through the church than our national political organizations could."

37 Ibid.

38 Ibid., 2.

One of the most obvious of "curses" to stem from the cohabitation of Saxons and other ethnicities was the very propensity for and increasing proliferation of mixed marriages. Indeed, Csallner believed the degenerative influences unleashed by such race-hygienic aberrations to be so seminal that "if we really want to, and still have enough moral strength to do so, then we will overcome mixed marriages—and if we do not surmount them, we will be destroyed."[39] While one can clearly see an increasingly impatient Csallner distraught with the sluggish translation of the Self-Help's race-hygienic ideology into policy within the church hierarchy, Csallner's language become outright militant in his fascist zeal when he demanded:

> If we are serious about our renewal, if we want to finally be holy serious [*heiligen Ernst*] about our National Socialism, then we can overcome the danger posed by mixed marriages—like virtually all other dangers—even if we do not have all the means of enforcement at our disposal that a state does. We can enlighten and educate ourselves and our youth. But because we have largely not been raised well enough, and haven't raised ourselves well enough, nor will we educate ourselves well enough to give others the great courage of responsibility—that demands the unequivocal willingness to act [*tatenfroh*] and self-sacrifice for the national community—that we need to also render the mixed marriage impossible, we might at least be able to educate our youth to educate itself to that end. And we can also battle the mixed marriages through proper career counselling, job provision [*Arbeitsvermittlung*], and the skillful steering of migrations.[40]

Returning to the familiar threat of exclusion from all social and political organizations, Csallner was to concoct increasingly draconian measures against undeterred Saxons endangering their nation's health. As the previously mentioned open letter to an unnamed Saxon woman illustrated, Csallner was adamant about the need to quarantine and expunge those who chose to defy this national imperative. Therefore it seems a natural progression when Csallner is found demanding that not only the perpetrators be excluded from all church and political posts, but that their parents be so as well, for having failed so miserably to instill a proper sense of national pride

39 Csallner, "Zur Frage der Mischehe," 10.

40 Ibid., 9.

in their children: "But not only s/he who has married a foreigner may not be given a honorary post or other power in our national (and church!) life, but also those who have failed to raise their children who have been able to do so. Besides, our church needs to realize that it is not just a Protestant church, but also a German one."[41] That said, it was only a short leap unto the conclusion that the church should quite simply refuse to wed mixed marriages, as it was absurd to confer God's blessing on something that so clearly threatened the Saxon nation's very survival.[42] Nonetheless, and in a further attempt to curtail mixed marriages along with any other deemed undesirable, Csallner established a "marriage counselling" office to discourage mixed marriages within the LSS's National Betterment (Volkspflege) department, which was also going to issue another (never realized) hallmark eugenic measure: mandatory "marriage suitability certificates" (*Ehetauglichkeitszeugnisse*).

Reviewing this chronological assembly of eugenic population policies that Csallner strived to implement through the church, and the rather brief time to have passed between them, the accelerating radicalization of his race-hygienic discourse is apparent. So Csallner was clearly exuberant when in July 1938 the "movement for renewal" took control of the church's constitutional assembly, the Landeskonsistorium.[43] Naturally, this landmark moment in the history of the Saxon church is of essential importance to Csallner's pursuit of his eugenic agenda through its infrastructure rather than via private initiatives or even his LSS. It is hardly surprising that Csallner subsequently resubmitted a number of his proposals, assuming that the ideological climate had become more conducive to his plans—so, for example, he tried to revive his failed 1927 effort via Siegmund's Fürsorgeausschuss to have the LK set its priests and teachers minimum knowledge requirements regarding racial hygiene and hereditary biology.

Regardless of the power politics which shook the Saxon church to its core in 1938, Csallner became increasingly bitter and disdainful of the Christian morality posited by many a local priest in opposition to fascist eugenics. If he claims the "working groups" he set up mostly through the LSS also served the purpose of exorcising the "theological nonsense" that distracted local priests

41 Ibid., 10.

42 Ibid. The church "must also forbid the consecration of mixed marriages, and hence prohibit that God's blessing is begged for [herabgefleht] what is a crime against our Germanness, and is so dangerous that we may perhaps by destroyed by it."

43 For example, Csallner, *Die volksbiologische Forschung*, 92.

from their national duties,[44] then Csallner's continuous tracts against conservative church dogma—which he thought retarded the onset and progress of the necessary regenerative efforts—became increasingly impatient.

In his 1939 piece *Irrwege* (where *irren* is to be wrong, and *wege* are paths, the most faithful translation would probably be *On Wrong Paths*), Csallner set out to analyze and substantiate his understanding of the role that the Saxon Protestant church must assume in the increasingly pan-German approach of Saxon politics to all of Romania's various German ethnic enclaves. Csallner was very much at the vanguard of the Saxon political self-perception as the natural leaders of this conglomerate, and while his own personal work focused almost exclusively on the various Transylvanian Saxon settlement areas (the Nösnerland, in particular), his organizations had initially hoped to reach and influence all of the diverse German groups. That said, Csallner opened his piece with the statement that "it is unnecessary to expound at length that our Protestant church has historically been and currently is the central support pillar and quintessential bastion of our German culture, first to us Saxons, but then also for all of the other Protestant, as well as for all the non-Protestant Germans in this country."[45]

Insisting that a "false religious morality" threatened to alienate and undermine people's connection with the church itself,[46] he proceeded to outline a host of faith-dictated practices he deemed detrimental to his eugenic cause. Csallner lamented the retardation of his agenda at the hands of church-led volunteer "work camps" (*Arbeitslager*) and the sermons delivered by its priests ("no, you are not becoming more capable, but less capable to serve our nation, and are damaging it by your deeds"[47]), and complained about the manner in which religious studies are taught in schools, avidly critiquing the inclusion of Genesis in its curriculum during the era "of German renewal, of the great self-awakening of the German nation!"[48] Csallner then turned to the church's reservations about eugenic population policies, to the religious notion that God had entrusted his greatest deeds not to the talented but the weak, that the church had failed to appreciate the Third

44 Csallner, "Das Landesamt," 19.
45 Csallner, *Irrwege*, 3.
46 Ibid., 5.
47 Ibid., 14.
48 Ibid., 15.

Reich's sterilization laws—exclaiming that "we had not even expected the church's civil servants to go to the trouble of acknowledging which categories of diseased the National Socialist state excludes from the German nation's hereditary future."[49]

Given this frontal assault on the core values of the Saxon church, Csallner arrived at the astonishing conclusion that "the church's pronouncements have become considerably more crackbrained and intolerant, and the friction between it and what can be asked of those somewhat capable of a healthy and educated mind has increasingly become unbearable—so that we are indeed facing the possibility of an internal split, that it may well come to pass that large sections of our youth in particular are not only entirely alienated by the church, but face it with hatred and think it little more than a temporarily necessary evil."[50]

There is, of course, a certain irony in Csallner's raving critique of the church, in that he had himself endeavored to employ it as a tool, a means to eugenic ends. He could hardly have conducted more than a handful of his numerous surveys and studies without the church's help and its substantial networks and archives. But as Csallner became increasingly frustrated with the church's rejection of his eugenic policy proposals, Fritz Fabritius's Self-Help came to offer him a second, more appreciative avenue through which to pursue his goals.

In 1938, Csallner published a chapter entitled "The *Auslandsdeutschtum* and the Racial Hygiene and Hereditary Health of the Reich" in the VDA's yearbook, which was reprinted two years later in *Volk und Heimat*.[51] It is one of the few articles (published outside of *Selbsthilfe* or *Ostdeutsche Beobachter*) to endorse National Socialism explicitly as a source of inspiration, while exhibiting a clear jealousy of the Reich's legislative and coercive capabilities to implement eugenic policies. Similarly, Csallner's consternation with the church's lackluster engagement with eugenics led him to conclude that "teaching the nation to see, think, and act according to biological laws is more difficult in areas where the church has a larger, perhaps even definitive influence, and is unfamiliar with the concept of hereditary

49 Ibid., 18.

50 Ibid., 20.

51 Csallner, "Das Auslandsdeutschtum."

health."[52] Returning to his usual mantra of decrying declining fertility rates overall, and the negative hereditary selection patterns represented therein, Csallner repeatedly insisted that Saxon fertility trends amounted to a negative selection.

As for the church, Csallner reiterated his previous demands that it restructure its wage and tax system, and "only allow such marriages that have been deemed racially and hereditarily unobjectionable by an approved doctor. Many couples might abstain from seeking the church's blessing; one or two may even leave the church. But a good national and church leader should be able to convince the wider masses that a responsible church cannot put up with, and certainly not bless that from which in all likelihood will come nothing but havoc and curses."[53]

III. Going It Alone: The Society of Child Enthusiasts, 1927–30

In the same manner in which we see Csallner's eugenic and population policies becoming increasingly dogmatic and totalitarian, so too with his organizations. Starting with the short-lived Society of Child Enthusiasts (or VdK) that aspired to create something of a *Gesinnungsgemeinschaft* through lobby work and the advancement of collective saving accounts, Csallner's ambitions were substantially boosted when after years of collaborating through Siegmund's Welfare Committee, Fabritius welcomed his idea of setting up and running a Self-Help Race Office in 1932. While its later permutations invariably became a pillar of Saxon population policy, Schunn is already found marveling in 1936 that Csallner's research had "a pivotal role for all of our other work [...]. We want to learn from it where our national body is damaged, the nature of the damage, and how to alleviate it. This office will hence be decisive in determining the entire future deployment of our energies."[54]

All began rather humbly, with Csallner's Society of Child Enthusiasts. Following a public lecture series spanning Schässburg, Mediasch, Gross-Schenk, Hermannstadt, and Bistritz in late 1926 and early 1927, Csallner founded the Society of Child Enthusiasts, his "movement" with himself as its leader (*Obmann*), at the inaugural meeting on May 18, 1927, in Her-

52 Ibid., 12.

53 Ibid., 13–4.

54 Wolff, *Ein Jahr Volksgemeinschaft*, 57.

mannstadt's Johannis Church.[55] It marked an important landmark in Csallner's eugenic career, while also embodying the core difference between his perception of the root cause of Saxon degeneration and Siegmund's. Rather than share Siegmund's insistence on a dwindling *Lebensraum* that necessarily resulted in his creation of the BSV, Csallner's society was to tackle the declining birthrate he so unequivocally deemed to be the cause and consequence of all of the Saxon nation's spiritual and racial ills. Inspired by the German Reichsbund der Kinderreichen, Csallner envied its growing "cultural movement" and ability to actively engage with legislative processes. But while the VdK was never to exert the degree of influence its German counterpart enjoyed, nor attain its membership, it marked Csallner's first institutional lobby group before eventually fizzling out three years later—only to be replaced by the markedly more aggressive and radical Rassenamt der Selbsthilfe, which Csallner founded in 1932.

That the VdK constituted a network of activists rather than a rallying point for the like-minded is evident, if nothing else, in its membership criteria. While Csallner must have been aware that it would limit the possible pool of candidates, applicants had to have at least four children (parents of three could be accepted if their marriage had not lasted ten years, widow/ers could apply if their marriage had lasted less than eight, or even less than five years if they had two children). In addition, the society's mission statement, its means and ends, are best described in a piece published both as a flier (*Flugschrift für Kinderfrohe*) and as an article in Siegmund's *Evangelischer Fürsorger*, under the telling title "What We Want."[56] Brimming with Csallner's usual worries about what continued reductions in Saxon fertility would entail ("if it continues as it has until now, if it does succeed to contain the child limitation in our nation—then the Germans here will die after all; then we are finished!"[57]), the article outlines the VdK's origins and underlying aims: "Oh, we are yet far too few who battle this plague, work without the necessary networks, and focus too much on acting through word and print alone. Barely anyone has tried to reach into life itself, to grasp and rework it. But to do precisely that throughout the Saxon lands, in various ways and in accordance with a cohesive plan, this is the aspi-

55 Heinrich Brandsch became his deputy, the bank clerk Hermann Büsch the society's legal advisor (*Erster Anwalt*), and Hugo Sitzler its treasurer (*Kassenwart*). This constituting meeting also set the annual membership fee at 60 lei, and the minimum required donation for patrons (*Förderer*) at 120 lei.

56 Csallner, "Was wir wollen."

57 Ibid., 9.

ration of the Society of Child Enthusiasts that hopes to slowly truly overcome our nation's suicidal child limitation from within."[58]

The Society of Child Enthusiasts had not been born of the economic needs plaguing large families as such, but primarily due to the need of "saving, of securing a future" for the Saxon nation: "We want to influence our nation's soul, its attitude to life and how to lead it, its sentiments—we want to bend its will."[59] Csallner thought the way to pursue this revolution from below was, surprisingly, by using the form and methodology of a political party that did not respond to public opinion, but created and furthered it: "We will, in most cases, only reach our goals if we—following a clear, considered, and consistent plan—work like a political party, if we become the 'public opinion,' and with our tenacious and undeterred work win over the powerful and influential who can petition or legislate or execute something, as well as the wider population; they may themselves realize this to be right and want it, or at least not prevent it through too much opposition."[60] Perhaps more ambitious (and unsuccessful) still, Csallner had hoped to create a pan-German society attracting members from all of the various German ethnic enclaves in Romania: "We call ourselves the Society of German Child Enthusiasts in Romania, and yet I always talk about the Transylvanian Saxons alone. Our movement is indeed currently limited to these, so that we should actually still call ourselves the Society of Transylvanian Saxon Child Enthusiasts. But we have chosen this name because we think it likely that our movement will sooner or later elicit similar movements amongst the other Germans of this country, and we count on the possibility that they will want to join us."[61]

The VdK's organ, albeit short lived, was the *Child Enthusiasts' Paper* (*Blätter für Kinderfrohe*), recast as the *Self-Help for Child Enthusiasts* (*Selbsthilfe für Kinderfrohe*) when the society officially became a Self-Help member in 1929. Initially an independent publication edited[62] and funded by Csallner,[63]

58 Ibid.

59 Ibid., 10.

60 Ibid.

61 Ibid., 11.

62 In addition to Csallner's articles printed in his *Kinderfrohe* range and discussed below, two further short but rather indicative pieces deserve a brief mention. The first accused the Saxon establishment of having allowed the nation's age pyramid to become horribly distorted by an ever-growing degree of family planning (Csallner, "Von einst und jetzt"), while the second investigated the "plague" of declining fertility rates in the Nösnerland unto 1905 (Csallner, "Die Kinderbeschränkung im Nösnergau").

63 Csallner's corrections to Ernst Wagner, *Laudatio auf Alfred Csallner*. Manuscript, June 1982, p. 1. Siebenbürgen-Institut Archiv, Ordner C, Alfred Csallner.

it was to become a *Selbsthilfe* supplement in 1930[64]—which was also the last year when it appeared. Apart from a few short articles penned by Csallner, the periodical also published longer passages taken from German authors, amongst which one finds rather interesting choices, such as Friedrich Burgdörfer's "Familie, Volk und Staat" and Günther Gründel's *Die Menschheit der Zukunft*.[65] Other tracts intended to instruct the reader in "biopolitical" thinking were the concluding passages of Albert Niedermeyer's article on "Die Minderwertigkeit und ihre Bekämpfung" (quoted from *Bundesblatt für den Reichsbund der Kinderreichen Deutschlands*, 1929) and excerpts from Hermann Werner Siemens's *Grundzüge der Vererbungslehre*.[66]

Apart from the lobby work and the numerous policy proposals Csallner unabatedly submitted to the church, the VdK's main goal was the creation of local *Spargemeinschaften*, or saving collectives that could subsidize its members with grants or low-interest loans. Csallner had founded his flagship local branch of the VdK in his local parish Stolzenburg, along with a "savers' group," on October 10, 1929, and boasted to have doubled its membership from nine to eighteen within half a year (while saving up 30,000 lei) in a remarkably emphatic 1930 article published by Siegmund's *Evangelischer Fürsorger*.[67] While the society regrettably has not left an archive for posterity, this article merits a closer look, as it yields a wealth of information on how the VdK was supposed to operate ideally. It quotes extensively from the society's statutes, which mostly pertain to the mechanisms through which funds were allocated to its members. Both the volume and conditions under which they are awarded were contingent upon the family's size and whether any of its members drank or smoked. Unsurprisingly, the first article emphasizes that "such members that do not smoke or drink spirits are, when it comes to calculating the level of subsidies and repayment instalments, awarded an additional child." Reflecting the educational mission Csallner assigned the VdK, its younger members would also be encouraged to abstain from alcohol and tobacco, by awarding them

64 Hauptleitung des VdK, "Zum neuen Anfang," 1: "The Self-Help has asked us to publish our *Blätter* as a supplement in their journal. We are most grateful for this proposal, and happily believe that the VdK's co-operation with the Self-Help will be to our nation's blessing."

65 *Selbsthilfe der Kinderfrohen* 4 (1930): 14–6.

66 *Blätter für Kinderfrohe* 12 (1929): 7–8.

67 Csallner, "Spargemeinschaften Gründen!"

an additional year's worth of membership fees.[68] Csallner, as always, concluded his piece with a euphoric call to arms: "Who has enough strength and courage to sweep others along with him? Write to me, get in touch, so that we may know and find one another and work together towards healing and blessing our nation!"[69]

Despite such enigmatic lobby work, the VdK failed to attract the volume of membership Csallner had hoped for, and joined Fritz Fabritius's Self-Help movement in 1929. Justifying this unambiguous shift in its agenda, Csallner publicly argued it had been motivated by socioeconomic rather than political factors:

> And if now ever-wider circles in our nation eagerly grab for the Self-Help that, founded and directed by our *Raiffeisenrevisor* and retired cavalry captain Fritz Fabritius, quickly spreads from Hermannstadt into the other parts of the Saxon lands, so it is again the large families amongst us that have all the reason to be joyous, no matter whether they themselves will use the Self-Help or just know that other large families in need can receive so incomparably cheap money.
>
> I don't even want to talk about all the great moral forces living in the Self-Help, and how through these our entire nation can be educated and morally elevated, and will be raised and elevated, but I just want to show how cheap the money that the Self-Help offers its members is.[70]

Although the society officially joined Fabritius's movement due to the "extraordinary advantages"[71] that the membership entailed fiscally, Csallner was himself consumed by the fascist project to regenerate and refashion the Saxon nation—to design and build a eugenic fortress bristling with a surplus of racially superior Saxon children.[72] And indeed, we find Csallner listed as the movement's representative (*Vertrauensmann*) in Roseln, where he became the village priest after leaving Stolzenburg utterly embattled, by July that year.[73] The VdK had consistently fallen short of his expecta-

68 Ibid., 15.

69 Ibid.

70 Csallner, "'Selbsthilfe,'" 8.

71 Ibid., 10.

72 As his friend Wilhelm Klein put it: "It was unavoidable that Csallner would exuberantly join Fritz Fabritius's movement, through which he hoped to succeed in his ambition to increase the Saxon population. He was an eager participant." See Klein, ". . . lies Kirchenteppiche pfänden."

73 *Selbsthilfe*, July 20, 1929.

tions, and Csallner, as he later recalled, was to "let it fall asleep" again before approaching Fabritius with the offer of founding and directing a Self-Help Race Office in 1932.

IV. The Self-Help Race Office, 1932–35

Csallner's ravenous embrace of Saxon fascism, which he was to help mold and translate into an empowered regime, was beyond doubt the single most radical reorientation (if certainly not revision) of his eugenic agenda. The Self-Help offered Csallner a politically, rather than clerically grounded framework within which to operate, and allowed him to establish the ever-evolving Race Office that underwent several drastic expansions in its rather short existence prior to being dissolved by Andreas Schmidt in 1943. Ominously absent from its 1975 version, Csallner's 1944 manuscript *My Scientific Work* expounds how and why he transferred his activities from the church to a centralized political organization, when he writes that his article on the church's *Irrwege* had only been part of a "larger project on 'Potential Population Policy Measures in the Transylvanian Saxon Protestant church' [...]. The other parts have remained uncompleted as—especially due to the strengthening of the National Socialist movement for renewal—ever more tasks that previously our church alone had seemed able to deal with can now be tackled by our national political organization, and as after 1941 the new national leadership, to our national detriment, has destroyed many of the church's key means of acting."[74]

Foreshadowing the rift that so effectively amputated his eugenic discourse in 1940, and although Csallner was to propose that several eugenic population policies should be micromanaged by the church (such as his plan for "hereditary leases"), this passage underlines the extent to which the resurgent and fascist Saxon political superstructure was to become his preferred means to eugenic ends. Csallner was to be involved in a myriad of population policies (such as honing the *Nachbarschaften*'s honorary gifts), but the rest of his brief but all the more industrious career was to be dominated by what started as the Self-Help Race Office.

74 Csallner, *Meine wissenschaftlichen Arbeiten* [1944], 15.

While Csallner's Race Office remained a rather virtual institution prior to its first expansion in 1935, it marked a triumph for Csallner by creating a political and institutional base through which he could pursue his eugenic agenda. His almost unbound enthusiasm for Fabritius—and for a movement flitting on the threshold of political empowerment it was to cross a year later—flows from the announcement of the office's creation, which Csallner published in the *Selbsthilfe* in October 1932:

> The very best that we, as a nation, own—and which forms the very precondition for a continuously industrious nation worthy of life and survival—is our hereditary substance. Our racial worth has, naturally, steadily declined since our immigration as a result of an undesirable degree of mixing with other nations; the emigration of our hereditarily best endowed; a declining fertility amongst our best classes; alcoholism and other processes, and perhaps this decline was never as rapid as it currently is. This is why we must act now to not only mobilize and preserve our hereditary fitness from further rot, but not least through the acquisition and securing of further *Lebensraum*, to encourage an above-average level of population growth, so as to turn an ongoing degeneration into regeneration. The latter goal will always be Nordification [*Aufnordung*]. But considering the particular realities within and outside of our nation, it might currently be prudent to not necessarily underline this point, and focus on the immediate goal of protecting and increasing that in our nation which is worthy.[75]

Not only does this proclamation provide a concise summary of Csallner's overall eugenic agenda, driven by the fear of a racial degradation stemming primarily from the failure of the most talented sections of society failure to reproduce their superiority at the desired rate, but it underscores the necessity of the re-homogenizing race-hygienic measures so fundamental to fascist eugenics. In a further dismissal of his traditional route to population policies—the church—Csallner proceeded to exclaim that "the Self-Help is now the only organization amongst us with the will and power to begin this task. It has taken up the work and entrusted it to a special Race Office under which, given the circumstances, further individual regional

75 Csallner, "Zur Errichtung des Rassenamtes," 4. For a full translation of Csallner's article, see Georgescu, "Transylvanian Saxon Eugenics."

and local race offices may be subsumed."[76] This boundless ambition notwithstanding, Csallner realized that the geopolitical climate of the early 1930s (essentially, prior to the ascendancy of the Third Reich) demanded a cautious approach. And so Csallner set his new office more realistic short-term targets, such as the organization of lecture series and the increased proliferation of race-hygienic knowledge. His primary target was to drive forward a sweeping assessment of the nation's racial health, so that the Race Office's "key task will hence be to study, town by town: our racial composition and our physical and spiritual merits in general, by especially highlighting the valuable and most valuable families; the spread of tuberculosis, venereal diseases, idiocy, epilepsy, alcoholism; the spread of family limitation; vital space [*Lebensraum*]; migration; debts; the progress of foreign nations [*fremde Volkstümer*]; protective and defensive options against imminent dangers; and more along these lines."[77] This ambitiously meticulous study of the overall Saxon constitution would be carried out, significantly, by "voluntary work groups," in particular by various Self-Help work teams (*Selbsthilfe Arbeitsmannschaften*, or SAM). Created as a superstructure for the Self-Help's youth movement, the SAM's involvement would, apart from the manpower, simultaneously offer the best opportunity to reach the masses, "to sharpen a sense of national responsibility in general, and their hereditary-health and race-hygienic conscience in particular, and thereby prevent many a calamity."[78] In other words, Csallner embraced the one ideology the organizational means and manpower of which had begun to rival that of the church: Saxon fascism.

v. The Reinvention of the Race Office as National Department for Statistics, Population Policy, and Genealogy, 1935–38

Shortly after Fabritius was elected president of the Association of Germans in Romania in June 1935, Csallner's Race Office was drastically expanded and rebranded as the National Department for Statistics, Population Policy, and Genealogy (LSBS). In November 1935, Csallner produced an eight-page-long pamphlet outlining its new mission statement and revamped organizational

76 Csallner, "Zur Errichtung des Rassenamtes," 4.
77 Ibid.
78 Ibid.

hierarchy, which assigned the LSBS its singularly most important task: a dauntingly ambitious survey of all German settlements in Romania, encompassing twenty-six individual questionnaires.[79] While this survey was bound to and indeed failed to meet the March 1938 deadline, the task was passed down to LSBS's successor, the National Office for Statistics and Genealogy (LSS).

Before excavating the little that is empirically known about this abortive survey, the preparation and logistics for which consumed the largest part of Csallner and the LSBS's time and commitment, it is instructive to ask how was the expanded National Department structured and how, if at all, did its mission statement change over time? The LSBS was headed by a national office whose director was appointed by the national leader (*Landesobmann*). Csallner, in his capacity as director, had an advisory seat on the German National Council (Volksrat), while being directly responsible to Wilhelm Schunn as the Commissioner of the Nation's Organic Constitution. The central office, for its part, appointed the leaders of its various regional (*Gau*), district (*Kreis*), and local (*Ortsstellen*) offices, while the latter were only set up in either particularly large or endangered settlements. On a final, structural note, the LSBS was funded (although most positions were honorary) by the National Council, as well as through local contributions.[80]

The LSBS's remit was substantially enhanced by the new brief it received in 1935. Conceding that the Saxons did not enjoy the luxury of a centralized state to enforce its policy decisions, and had to rely on consensus-building instead (or, rather, peer pressure), expanding its propagandistic remit was a central ambition. Therefore it is not surprising to find Csallner adamantly arguing that the LSBS must be "an office that, above all else, must train speakers [*Aufklärer*] and recruiters, must get and provide all of the material they need for their enlightening and publicizing work.[81] It needs to be an office that always knows exactly, to the minutest detail, what the state of our *Lebensraum* and hereditary composition is, as well as what possibilities to grasp when they offer themselves."[82]

79 Csallner, *Die Arbeitsstellen für Statistik, Bevölkerungspolitik und Sippenwesen.*

80 Ibid., 6–8.

81 So, for example, Csallner had already acquired over a thousand copies of the pamphlets published by the Reich Committee for National Health Services (Reichsausschusses für Volksgesundheitsdienst), and had succeeded in securing a number of copies of the journals *Neues Volk* and *Archiv für Bevölkerungswissenschaft (Volkskunde) und Bevölkerungspolitik* for "a large number" of his coworkers and for the LSBS's library. Ibid., 7.

82 Ibid., 3.

Naturally, one of the centerpieces of Csallner's attempt to carry his brand of eugenics and its particular understanding of heredity into every household came in the guise of genealogical research. He hoped to encourage everyone to study their particular ancestry, to have them identify that which was desirable (and what less so) in their hereditary makeup. Csallner had frequently defended the various virtuous fruit the genealogical tree could offer in terms of deciding whom (if at all) one should marry and have (at least three) children with; what profession had been hardwired; and to awaken the nation to its blood ties (*Blutsbande*) as they formed through internal migrations. Genealogy was thus something of a race-anthropological microcosm, an easily accessed window into the historical past, but also a Janus head predicting the future, as "hardly anything is so suitable to permanently enthuse so many people, to let them recognize the laws of heredity and the importance of hereditary traits by themselves, to teach them how to see and think biologically, to awaken in them the sense of community and a communal destiny, and their responsibilities towards each other."[83] However, the proliferating genealogical research amongst the masses was not an end unto itself, but also integral to Csallner's sacralization of statistics, and his desire to pool these thousands of family trees in a vast database that could rival those of any "modern state." The beating heart of this new database, predicting and prescribing virtually everything from the cradle to the grave, was the "hereditary-health and fitness archive" (*Erbgesundheits- und Erbtüchtigkeitsarchiv*), which he aspired to create in union with the medical establishment and the church.[84]

Thus, Csallner had set out an ambitious agenda, and expected all of the regional and local branches to be operational by the end of 1936 in order to begin the equally ambitious census designed to be conducted by Schunn's neighborhoods in all the German settlement areas concurrently and in secret in late 1937 (and which he thought would only take two or three months to complete, with provisional results ready to be presented as early as 1938).[85] The very idea that such an endeavor could be pulled off without the interference of the Romanian state and in spite of the internal political upheav-

83 Ibid.

84 Ibid., 3–4.

85 See Viktor Quandt's memo to all neighborhood officials from January 1937: NA Sibiu, CGR, D.XIII.B.477, p. 36. Also see Csallner's memo from January 20, 1937, regarding the completion of his training courses for neighborhood officials, and the additional guidelines on completing the survey's questionnaires: NA Sibiu, CGR, D.XIII.B.562, p. 90.

als that followed Bonfert's failed coup was almost naïve given the extensive data Csallner wanted to collect.[86] This was still the case even with Fritz Fabritius's direct order to all national and local offices of the NAF to cooperate and assist in its successful completion.[87]

As was to be expected, Csallner's census was quickly brought to a standstill by the inner political turmoil dividing Romania's Germans along increasingly polarized picket fences, in addition to the state's dislike and suspicion towards this mammoth undertaking. Csallner was forced to suspend his study in the wake of arrests, interrogations, and harassment "endangering" his students' fieldwork, which culminated in the confiscation of three meters' worth of material by the police.[88] As he explained two years later in a memo on the LSBS's expansion into the LSS, Csallner had been aware of the risks but had not expected the level of subversion and hostility that it did indeed provoke:

> It had been a risk to begin such a thorough, family-by-family survey of 800,000 national brothers, living in far more than 1,000 settlements, with a national organization that was itself only being built, and that had also been weakened and paralyzed by internal battles. All the more so as one initially did not even have the appropriate staff, had to train them in dozens of workshops, and anyway also knew in advance that the work of these hundreds, indeed thousands of coworkers, regardless of the fact that it was neither opposed nor dangerous to the state [*Staatsfeindlich oder gar Staatsgefährlich*], would nonetheless have to be done in secret, should they not run the risk of being inhibited or even broken up by the state. Then forces emerged against this census that were considerably graver than I had expected. People loyal to the national community, especially due to the repercussions of battling the DVR, were not always as readily available to me as I had hoped, and members of the DVR not only frequently refused to give the information asked for, but sought to agitate against our survey and make our work more difficult. But the state authorities in particular noticed some of our work, which led to numerous house searches and interrogations—yes, even led to arrests.[89]

86 For the details of all the twenty-six questionnaires comprising the census, see Csallner, *Bestandsaufnahme*. See also Csallner, *Anleitung zur Bearbeitung der Fragebögen*.

87 Fabritius to NAF, November 14, 1936, NA Sibiu, CGR, D.XIV.840, p. 5.

88 Csallner, *Meine wissenschaftlichen Arbeiten* [1975], 18–19; and Csallner, "Die rumäniendeutsche Bestandsaufnahmen," 216.

89 Csallner, "Das Landesamt," 3.

Although an enormous chunk of his data had been taken by state authorities (particularly that of one Bessarabian and twelve Banat districts), Csallner managed to save about half of his material on Bessarabia, that on about seventy settlements in the Banat, and in addition to several rural communities, had salvaged the data for the cities Kronstadt, Schässburg, Mediasch, Hermannstadt, Bistritz, and Reps.[90]

VI. The National Office for Statistics and Genealogy and Its Six Departments, 1938–41

The LSS, from its humble origins as a Race Office in 1932 to the radical expansion of 1938, was the institutional manifestation of Csallner's perception of a national racial dysgenic crisis, and, ultimately, the means to counter it. It marked a clear departure from working solely within the organizational framework afforded by the church, and created a research institution both informing and increasingly making policy decisions: "The National Office for Statistics and Genealogy does not merely endeavor to evaluate all statistical matters concerning the German national community in Romania, to alert them to their national duties, and to guard their appropriate application, but what is more, has the duty to ensure that all other national and race-hygienic duties are realized and properly implemented by any other relevant departments [*Stellen*]."[91] The LSS was intrinsically linked to Csallner, and he headed both the national office in Hermannstadt and the Transylvanian regional office, "in order to be able to work in at least one *Gau* myself—to not only stimulate the work in others through central directives, but to influence and drive forward by example."[92]

Of the handful of publications and manuscripts pertaining to the LSS's inner workings (currently known) to have survived, the most significant is a twenty-seven-page-long report by Csallner from October 1939.[93] This report, along with the appended questionnaires, graphs, and maps, constitutes the documentary core of any substantive analysis of the LSS's agenda and activ-

90 The National Archives in Sibiu hold the following of what is left of this data: CGR.D.XIV.850: Bessarabia: folders 689–730; Bistritz District: folders 731–43; Braşov District: folders. 744–65; and Sibiu District: folders 766–808.

91 Csallner, "Das Landesamt," 1. See also Csallner, *Organisationsrichtlinien*, 3.

92 Csallner, "Das Landesamt," 1.

93 Ibid.

ities in the latter half of the 1930s. Regrettably though, Csallner does not provide much detail on its employees or their respective roles within the organization, nor does he elaborate on their research interests in any meaningful detail (while the report does include a very interesting four-page-long lament of his fiscal woes and pleas for more funds). Instead, the manuscript offers a rather detailed structural analysis of the LSS's central mission statement(s) in the wake of its recent expansion, and an outline of the work it would (but was largely unable to) carry out in the following years. Nonetheless, taking a closer look at its departments and at the briefs Csallner had composed for them provides an intriguing insight into his overall eugenic agenda, and on the collaboration it entailed with Schunn's neighborhoods.

The LSS added a number of new departments (all of which included a varying number of individual subsections) to the existing infrastructure established by its predecessor, and by 1939 consisted of six departments tasked with Statistics (*Statistik*); Genealogy (*Sippenwesen*); National Betterment (*Volkspflege*); *Lebensraum*; Foreign Nations (*Fremde Völker*); and Enlightenment, Recruitment, and Education (*Aufklärung, Werbung und Erziehung*).

a. Statistics

The Department of Statistics was tasked "with collating, evaluating, and providing all necessary documents to the national leadership, and to work towards an increasingly perfected meshing together of all *völkisch* organizations."[94] Its main functions were the accumulation and evaluation of demographic data, assessing the extent of the dysgenic crisis, and devising solutions for it.

Of its six individual sections, the first and most industrious one was Censuses (*Bestandsaufnahme*). It was given the immediate responsibility of completing and evaluating the data left from Csallner's abortive 1937–38 census, augmenting it with further surveys where needed. Under Csallner's direct supervision, Transylvania's regional data was to be the only set evaluated by late 1939. While the Bukovina material had been fragmentary, its regional director (*Gauwalter*), Dr. Herbert Mayer, subsequently conducted his own survey to complement it. But to Csallner's consternation, Bessara-

94 Ibid., 2.

bia's regional director, Prof. Arthur Fiechtner, had assumed a teaching position in Radautz, Northern Bukovina, which prevented him from benefiting from the VDA's provision of funds for a ten-month stay in Hermannstadt to evaluate the remaining data sets.[95] Instead, Csallner was to rely on the foreign students who had initially aided him in 1937, and sent the bulk of the material to Vienna, to be analyzed there.[96]

A second section, and one of Csallner's flagship ventures, was the recently created Neighborhood Index File (*Nachbarschaftskartei*), which Wilhelm Schunn had asked the LSS to design and manage. The system Csallner devised was a file index system that clustered data on individual families (or *Sippen*), and was geared towards the annual evaluation of each neighborhood's members (with the file itself renewed every ten years). But above and beyond the mere collection and maintenance of personal data, the index was to offer something of an up-to-date barometer of national demographic and economic trends (particularly so in relation to career counselling). While Csallner hoped to convince Schunn of the need for a "drastically enhanced index," the existing version already embodied a remarkably detailed centralized database. Its function was not limited to the surveillance of economic and demographic trends, but it offered the empirical foundation for political interventions "to prevent the overcrowding of particular jobs, and hence simultaneously prevent our comrades from immigrating to the Old Kingdom."[97]

The degree of total statistical control Csallner strived toward is perhaps more apparent here than in any other text when he writes that: "In a few years' time we will be able to evaluate the statistical data for all Germans annually, to do so more thoroughly and frequently than generally most states can, and will simultaneously create the most powerfully compelling means imaginable to quickly and purposefully intervene in its development."[98] Clearly, Csallner's quest to amass increasingly vast databases (such as his 1937–38 census) stemmed from a sacralization of statistics as the ultimate

95 Ibid., 3.

96 Although Csallner does not specify to whom in Vienna he sent this material, it could well be Dr. Josef Kallbrunner, with whom he was already collaborating along with the DAI towards copying and evaluating church archives in the Banat as a basis for that LSS regional office's work. On this, see the letter from Nikolaus Hans Hockl to Mr Zilliken dated October 28, 1941, BA Berlin, R 1509, 1622.

97 Csallner, "Das Landesamt," 4.

98 Ibid.

in objective self-reflective narratives on societal (d)evolution, and simultaneously the objective determinant of centrally implemented eugenic population policies.

This desire to collect ever more data is also evident in the remit Csallner assigned to a third section, Special Surveys and Advisory Services (*Sondererhebungen und Beratungen*). Amongst others, we find seven indicative surveys either being prepared or already underway in 1938. The first, on "Landownership in Saxon Villages" ("Hofbesitz in den sächsischen Dörfern"), was being carried out by a group of student volunteers who worked alongside two of Csallner's working groups of teachers and priests, and focused on the Bistritz county parish.[99] A second survey dealt with the related question of "Urban House Ownership" ("Städtischer Hausbesitz"), and was due to be overseen by Schunn's neighborhoods. A third, on "Mixed Marriages," eventually formed the empirical basis for Csallner's three-part article thereon, while simultaneously he had also begun preparatory work on extending his study to include the Bukovina (which he deemed particularly prone to interethnic marriages) in 1940. A fourth survey dealt with "Migrations" ("Wanderungen"), and informed the article on the prevalent migratory trends in the village Grossalisch discussed earlier. In addition, a fifth census studied the "Rural Youth" ("Ländische Jugend") in conjunction with a sixth, on the critical issue of employment patterns between Saxon and non-Saxon "Farm-hands" ("Knechte und Mägde"). Not only was Csallner alarmed to note the increasing frequency with which Saxon youths (in particular women) emigrated to the Romanian Old Kingdom, he also returned to his customary complaint that employing anyone other than a Saxon would by definition amount to a self-destructive loss in nation wealth. Most of the ethnically Romanian servants in Saxon employ tended to live themselves in Saxon settlements, which inevitably led to an ongoing loss of *Lebensraum*, as "with these hundreds of millions of lei which we pay Romanians year on year—to our most dangerous adversaries in the struggle for our continued existence—we strengthen them precisely

99 Ibid., 5: "I have conducted these primarily with students who offered their services to me as scientific helpers, first individually, then organized as a *Fachdienst*. Now I have also given two of the work groups of priests and teachers I set up the task of studying property ownership, and paying particular attention to the Bistritz church district, as its particularly high lack of children has led to a particularly high number of abandoned houses and farms, while for the same reason numerous property owners now hold more than one farm."

there where they have already breached our *Lebensraum* and are snatching up more of it, piece by piece."[100] The last of the seven surveys alluded to here was hence to preoccupy itself with the question of "Territorial Gains from, and Losses to Romanians" ("Bodenflächen gewonnen und verloren mit bezug auf Rumänen"), benchmarked for summer 1940. In addition to this ambitious workload, Csallner had also been entrusted with what certainly would have been a highly politicized survey quantifying the interwar "losses" that the German enclaves suffered at the hands of Romanian "purposeful overtaxation," the "agricultural reform's great land robbery," the decline in German civil servants, and so forth. The stated goal of this conglomerate of studies was hence to "establish the injustice that we have suffered in the past twenty years, and continue to suffer, and to empirically justify our demands for compensation."[101]

In equal measure enthralling and unnerving due to the absolute absence of any further documentary evidence for its activities, the section Extraordinary Surveys and Advisory Services also featured a Subsection on Racial Research (*Rassenforschung*). It was headed by Albert Hermann, who had already published a landmark biometric study on the Saxon peasantry in southeast Transylvania. Hermann's assistant was Michael Fleischer, who had carried out race-anthropological studies of Saxons in the northeastern regions of Transylvania with a research grant negotiated by Csallner. In short, this subsection was tasked with advancing the biometric study of all Saxon communities (and, interestingly enough, was to include local Romanian and Szekler populations in their studies), with the long-term goal of extending its remit to cover all German enclaves in Romania.

A fourth section operating within the Statistics Department dealt with foreign surveys (*fremde Arbeiten*) and had "the task of evaluating the statistical and related work conducted by others insofar as they are pertinent to the Germans of Romania; to adopt useful portions therein, to correct their errors, and battle that which is damaging and dangerous."[102] In particular, it was supposed to vet, amend, and correct data released by the state, such as the "fundamentally wrong" ethnographic map Emanoil Simtion produced

100 Ibid, 7.
101 Ibid. While this study appears to have remained incomplete, Csallner did produce an unpublished manuscript entitled "Damages Inflicted upon the German National Community in Romania by the State" ("Schädigungen der deutschen Volksgruppe in Rumänien durch den Staat") in 1940.
102 Ibid., 8.

using data from the Romanian census in 1930.

Many of the maps, graphs, and illustrations already encountered in this study were the work of Günther Herbert, a close friend of Csallner's and head of the fifth department, Representation and Visualization (*Darstellung*), while the sixth and final department to be housed in Statistics was the LSS's archive (*Sammlung*).

b. Genealogy

The LSS's second department was Genealogy (*Sippenwesen*), which was assigned the critically important task of "assessing the national community's genealogical [*sippenmässigen*] makeup, to teach it to think, see, and act in the genealogically and hereditarily [*sippenhaften und sippengerechten*] proper way. To that end it has the particular duty to preserve and evaluate parish registers, and to introduce and encourage the private pursuit of genealogical research and hereditary health."[103] The Genealogy Department was subsequently subdivided into five individual sections, where Documents (*Urkunden*) was briefed with the aggregation of "register excerpts and other documents, to certify *Ahnennachweise,* and is primarily used by those to have immigrated to Germany and by their descendants."[104]

A second section, headed by Father Gustav Arz, was dedicated to the study of parish registers (*Kirchenbücher*). Inspired by a visit to the Kirchenstelle Alt-Berlin, and with Bishop Glondys having consented to his design, the tedious work of mapping all parish registers was seminal to Csallner's ambition to compile genealogical studies not merely of individuals or their families (for which he could always rely on either *Nachbarschaftskartei* or *Erbtüchtigkeitsarchiv*), but of entire settlements. What is more, Csallner hoped it would serve to popularize the *Volkskörperforschung* that had faded into obscurity when Father Johann Bredt succumbed to his illness in 1936.[105]

That said, the third section, Genealogical Research (*Sippenforschung*), intended to further introduce and immerse the wider populace in the brand of "biologically framed" genealogical research that Csallner had been advo-

103 Ibid., 9. See also Csallner, *Organisationsrichtlinien des LSS*, 7.
104 Csallner, "Das Landesamt," 9.
105 Ibid., 10.

cating since the early 1920s.[106] This section was also to lobby the political and church authorities to coerce their employees into submitting their own family trees, and to thereby force the public at large to grapple with the concept of an immutable heredity at least temporarily, if not spark a permanent enthusiasm. Not surprisingly then, a fourth section constituted the LSS's genealogical archive (*Sippenarchiv*).

The fifth and final section housed in Genealogy was Heraldry and Symbols (*Wappen und Zeichen*). It was headed by the prominent heraldic expert Dr. Albert Arz von Straussenberg, who had been intimately involved with the design of the neighborhoods' insignia. Straussenberg's engagement with the LSS, though, had actually been rather accidental. He had initially approached the Cultural Institute in Hermannstadt with the proposal of setting up a heraldic department when Csallner offered him free reign in his LSS. Csallner had initially thought this section would do little more than collate information on already existing coats of arms, and act as an advisory body to those endeavoring to create new ones, but it quickly gained a considerably more significant role, which Csallner defined as follows:

> With the creation of the department Heraldry and Symbols I believe to come a good bit closer to realizing an old dream of mine: I would like to assist in the creation of a caste of German "noble leaders" [*Führeradel*] that will constitute a closed—while not too far removed from the rest of the community—"breeding community" [*Zuchtgemeinschaft*] raised to noble thinking and noble composure in the clear awareness of their particular duties towards the national body, namely to be and do all that it can in the coming centuries along the lines of what Hans F. K. Günther demanded of a *Führeradel* in his *Führeradel und Sippenpflege*.[107]

Csallner's embrace of Günther is also interesting because it alludes to the close affiliation of Saxon eugenics with the more extreme branches of German racial hygiene. These close ties were partially the result of the substantial number of Saxon eugenicists linked to Eugen Fischer specifically (such as Hermann, Bredt, and Hügel), as well as one of Siegmund's early influences.

106 Ibid.
107 Ibid., 11.

c. National Betterment

National Betterment (*Volkspflege*) was clearly going to be one of the LSS's most important departments. It was tasked with surveying the past, present, and prospects of the Saxon hereditary stock and, in close cooperation with the National Office for National Health (Landesamt für Volksgesundheit), to "awaken ever wider sections of the population to their responsibility for the nation's future composition and fate, to awaken and entrench in them the pride and willingness to bear these responsibilities for the recovery and regeneration of the German people."[108] Despite the considerable degree of jurisdictional overlap Csallner created amongst the LSS's six departments, he explicitly tasked National Betterment and its five sections with the battle against small families; the threat of differential fertility; the nation's spiritual and racial ruin through mixed marriages; and the socioeconomic scourges, alcohol and tobacco.

The section Population Size (*Volkszahl*) was hence supposed to "primarily lead the battle against declining fertility rates and work towards a strong German population increase"[109] in close collaboration with the neighborhoods. Wanting to add to the incentive provided by the honorary gifts, Csallner introduced a "book of honor" with a matching "honorary emblem" (*Ehrenzeichen*) for their recipients. He also resubmitted his 1927 outline for the "equality fund" (*Ausgleichskasse*) to the church via LSS, insisting that "back then I found no appreciation [for my plan]. Now though the time has come, as members of the 'movement for renewal' form the decisive majorities in the church's corporate bodies, and the Landeskonsistorium in particular, and one can expect the next *Landeskirchenversammlung* to finally accept my or a similarly thorough solution."[110]

In an increasingly draconian mindset, Csallner tabled a considerably more radical policy paper proposing the exclusion of all who could, but failed to bear more than four children, from all church and political offices. In other words, he advised to effectively exile them from the Saxon community at large:

108 Csallner, *Organisationsrichtlinien des LSS*, 7–8.
109 Csallner, "Das Landesamt," 12.
110 Ibid.

> I am expecting particularly favorable results from a decision that I shall propose to the national leader, and which I will ask the Landeskonsistorium to present as a draft law to the next *Landeskirchenversammlung* [. . .]: those who, although they could, do not bring themselves to have at least four children—and are thereby refusing to give the national community the most valuable of all gifts—should henceforth be excluded from all national and church honorary positions, thereby taking away from them any means of partaking in national and church life.[111]

As "quantity" was only part of the equation, it fell to the section Heredity and Race (*Vererbung und Rasse*) "to ensure that beyond their mere number, their quality is not forgotten."[112] This section thus dealt with two of Csallner's greatest concerns: the continual loss of racial nobles due to particularly low birthrates amongst the hereditarily most valuable sections of society, and the loss of ethnic homogeneity: "It has to lead and strengthen the nation's resistance to mixed marriages by calling on one's national pride and emphasizing one's duty towards it. But it also has to discover the causes for mixed marriages, and work towards their resolution."[113]

The third section, Marriage Counselling (*Eheberatung*), was to actively promote the introduction of "marriage licenses" (*Ehetauglichkeitszeugnisse*): "I also believe that following the appropriate enlightenment and advertising/courtship, our Protestant church will, in the foreseeable future, only carry out church weddings if 'marriage licenses' are produced."[114] Csallner had floated this idea in an article published in Siegmund's *Evangelischer Fürsorger*, seeking to convince the church that these race-anthropological evaluations were both a national imperative and of great statistical utility: "Our church can and must, if it is aware of its responsibilities, make a church wedding conditional upon the submission of a marriage license conferred by doctors it trusts and of whom it can ask, as they will examine almost the entire maturing [*heranwachsende*] German population medically and hereditarily, to produce a report for the hereditary archive."[115]

A fourth section, founded in September 1939, led Csallner's battle against society's twin ills of alcohol and tobacco (*Weingeist und Tabak*). It is little

111 Ibid., 13.
112 Ibid.
113 Ibid.
114 Ibid., 14.
115 Csallner, "Familienbuch—Erbarchiv," 2.

surprise that he would have created such a section, considering Siegmund's influence and their roles in the IOGT. Its placement within the organization is very much in tune with Csallner's understanding of their negative impact lying less with the nation's actual health, and much more with its wallet: "Not only have I, for many years now, industriously labored in this field and want to continue the work that has grown close to my heart, but also because the greatest threats inherent to alcohol and tobacco consumption are not even health related, but constitute an eerie and irresponsible waste of national wealth in addition to severely maiming our moral [*sittlich*] strength to resist in the battle for our continued existence."[116]

A fifth section, Festivals and Their Organization (*Fest- und Feiergestaltung*), was set up to ensure that all festivities and festivals took the basic tenets of "national betterment" into account by celebrating the large family and its beneficial, revitalizing impulses.

d. Lebensraum

The fourth of the LSS's six departments, *Lebensraum*, subsumed three individual sections and was charged with assessing the extent of territorial losses to ethnic "others," and devising means to counter it. Its first section, Property (*Besitz*), was assigned the task of exploring rural settlements in relation to: quantifying the amount and size of abandoned properties (due to, for example, emigration or inheritance); the employment of "foreign" agricultural laborers and handymen; and the extent to which land holdings (and their individual plot size) were shrinking. It was also to promote the creation of local "work groups" by providing conceptual and methodological support.

The department was also supposed to prevent any further territorial degradation by flagging areas in need of Saxon human capital and advocating the importance of employing and selling only to one's ethnic kin. A typical example of what form this sort of encouragement was to take is afforded by one of LSS's most direct appeals, in this case to the "Saxons of Arkeden!" ("Sachsen von Arkeden!"). A case study of striking simplicity, the poster featured two pictures, where one was of the local church-castle and the latter of a street. Their respective captions read: "Your fathers have built this church-

116 Csallner, "Das Landesamt," 14.

castle so the community may brave the storms of history and persist unto today" and "What do you do to protect yourselves from those who threaten you? You let Gypsy hawks buy up this entire street!"

A second section, Career Counselling (*Lebensstellen*), had a rather difficult task considering that the very idea of a *Lebensstelle*—a lifelong, permanent position or job such as, for instance, a family-run pharmacy—was increasingly losing its economic significance. Nonetheless, feeling embattled by the perceived onslaught of ethnic others into the job market, the notion that this department must enlighten and encourage the masses to "not only defend and further the existing ones, but struggle to regain lost *Lebensstellen* while creating new, German ones"[117] offered a more nationalist than utilitarian form of career counselling.

The third section, covering migrations (*Wanderungen*), consequently extolled how internal migrations created and perpetually recreated an organic nation bound by blood and guided by a communal destiny (*Blut- und Schicksalsgemeinschaft*). Based on this propagandistic effort, the department encouraged migrants to settle in German rather than "ethnically foreign" regions, and led the battle against emigration more widely—mainly by focusing on "containing the frightful emigration to the Romanian Old Kingdom" and Germany.[118]

This section also featured a rather interesting subsection, if only in its Transylvanian Regional Office, namely Transylvanian Saxons Abroad (*Siebenbürger Sachsen in der Fremde*). Set up to oversee one of Csallner's favorite pet projects, it sought the current addresses of emigrated Saxons in order to send them "letters from home" (*Heimatbriefe*).

These tried to encourage the prodigal sons' return home, while tending to their "Germanness" in the foreign wilderness.[119] Ideally, Csallner wanted to publish four (about twelve-page-long) issues every year in collaboration with the VDA Kassel, and had asked the Saxon societies in Graz, Budapest, and Vienna for their membership lists in late 1938.[120] In fact, he had actually succeeded in amassing files on about 10,000 émigrés by the time Andreas Schmidt dismantled the LSS in 1943. Csallner complained bitterly when

117 Ibid., 15.
118 Ibid., 16.
119 Ibid., 17.
120 NA Sibiu, CGR, D.XIV.688, p. 13.

Schmidt destroyed large swaths of the archive instead of moving it to the national leadership's new base in Kronstadt, especially as he was convinced that Schmidt "had it pulped, regardless of how my daughter tried to save it, only to hurt me: 'If one were to need such a thing in ten or twenty years, one could simply do it again.'"[121]

e. Foreign Nations

Reflecting the heavy emphasis Csallner placed on preemptive population policies, the LSS's fifth department, Foreign Nations (*Fremde Völker*), monitored the degree and nature of interaction between Germans and their land-hungry "neighbors." Its task was "to observe and study the relationship between the German ethnic group(s) and the other nations [. . .], to pay special attention in particular to matters concerning culture, the economy, and biological trends [*Erscheinungen*]."[122]

Founded in early October 1939, it was also one of the LSS's newest innovations, and remained largely uncompleted. Methodologically, its departments were structured ethnically at the regional level (while the National Office only established a respective section if the same "ethnic other" was found in at least two regions). While we know that the subdepartments overseeing developments within the Jewish, Gypsy, and Armenian communities were operational by 1938, it is currently impossible to discern how many there eventually were, or even how they operated on the ground.

The exception is, of course, the department's activities in Transylvania. Its regional office housed sections on the Romanians, Szeklers, Gypsies, and Jews, and Csallner assigned it several telling surveys. For one, he was concerned with the potential detrimental impact that falling birthrates and debt levels amongst Romanians could have, and launched a study to investigate "the incredible biological damage incurred by an internally unprepared Romanian nation as a result of its sudden political and social rise since 1918."[123] A second survey was to deal with the "structural changes amongst

121 Csallner, *Meine wissenschaftlichen Arbeiten* [1975], 17. A significant portion of this data has survived, namely 281 pages of lists detailing each emigrant's name, profession, age (and date of birth), date of departure, address, as well as the names and contact details of close relatives in Transylvania. See NA Sibiu, CGR, D.XIV.846.

122 Csallner, "Das Landesamt," 17.

123 Ibid.

the Szeklers that have led to a dangerous degree of infiltration of our cities." Turning his attention to Gypsies, a third study was to assess what Csallner deemed to be their ruinous (*verderblich*) influence on Saxon morality, in the Reps region in particular, and to evaluate the damage suffered by the Saxon *Lebensraum* at their hands (especially with regards to the already encountered example of Arkeden). Finally, a fourth study was to investigate the perceived Jewish extortion (*Auswucherung*) of Saxons in Bistritz and Sächsisch-Regen.[124]

f. Enlightenment, Recruitment, and Education

The sixth and last LSS department was Enlightenment, Recruitment, and Education, which was to facilitate ideological indoctrination by "coordinating and furthering the informative, recruitment, and educational projects of the other departments," ensuring "their frictionless and increasingly complementarily augmenting of each other's [...] educational work."[125]

To that end, its six individual sections oversaw and "influenced" the press (*Presse*) in cooperation with the National Department of the Press and Propaganda (Landesamt für Presse und Propaganda); the section Libraries (*Büchereien*) managed and expanded the LSS's house library and set up "wandering libraries"; the third produced and proliferated educational pamphlets (*Aufklärungsschriften*);[126] the fourth trained speakers and organized lectures and seminars (*Vortragswesen*); the fifth acted as the curator of exhibitions (*Ausstellungswesen*); and the sixth section, Education and Training (*Unterrichtswesen*), was meant to readdress a Romanian school curriculum wholly unsuitable to raising "German children." In fact, Csallner had himself written teaching materials based on his research, and seems to have even had his textbooks field tested and approved by an unnamed school in Schässburg.[127]

124 Ibid.

125 Ibid., 18.

126 An interesting example of how Csallner continued to sustain and further deepen his relationship with the Reichsbund der Kinderreichen Deutschlands, which had given him the initial impetus to found his short-lived VdK, is that they sent him 200 copies of the *Völkischer Wille* (*National Will*), which Csallner, under the greatest of fiscal lamentations, claimed he was unable to post to his coworkers as he could not afford the stamps. Ibid., 23.

127 Ibid., 18.

As Csallner was consistently adamant about the need for further funds to translate these grand plans into practice, his November 1939 report on the LSS's restructuring also included a lengthy exposition of his precarious fiscal position. Offering a rather curious insight into the day-to-day inner workings of his organization, Csallner insisted that his small office was utterly "unworthy" and "undignified," as he shared it with his coworker Julie Henning. Csallner also complained that Günther Herbert, his "most important" coworker, was forced to work from home. Worse still, the lack of travel funds meant that Csallner had only been able to visit Bessarabia and Dobrudja once since the creation of the Race Office in 1932. His poverty, Csallner continued, had become so unbearable that he considered resigning from his honorary position as director of the Science Department (*Abteilung Wissenschaft*) of the Hermannstadt Cultural Office (*Kulturamt*), which he had assumed in March 1939, but that was a short-lived post dissolved by Andreas Schmidt and eventually replaced by the Forschungsinstitut der Volksgruppe, in which Csallner had no further role.[128]

Csallner also underlined his personal importance to the Saxon eugenic discourse, and that of his LSS as its nurse and caretaker. Truly unequivocal, Csallner insisted he had been instrumental in advancing Saxon race-hygienic research in addition to having reared a new generation of laymen-come-racial-anthropologists: "I can calmly claim that as yet no one has made as great an effort to focus our science on these true-to-life and nationally important tasks—while supplementing our current lack of proper researchers with an increasing number of laymen—as I am doing, and as I have done long before I was entrusted with the care for our country's German science."[129] Grounded in this self-assessment, Csallner moved to making a clear demand for a substantial salary hike, threatening that "despite all my eager willingness to serve my nation and give her my best, I will have to look for another salary, and at least lay down my full-time directorship of the National Office for Statistics and Genealogy."[130]

Although Csallner did not resign, his career as one of the most prominent Saxon eugenicists came to a rather abrupt end almost precisely a year later, in the wake of Andreas Schmidt's return from Berlin. Andreas

128 Csallner, *Meine wissenschaftlichen Arbeiten* [1975], 14; NA Sibiu, CGR, D.XIV.835, pp.17–18.
129 Csallner, "Das Landesamt," 25.
130 Ibid.

Schmidt's appointment as National Leader marked a critical shift in Saxon sociopolitical realities, one that did not move in Alfred Csallner's favor. Initially, he was delighted to see the LSS further expanded into the Institute for Statistics and Population Policy of the German National Community in Romania (ISB), which was also granted the larger budget he had demanded.[131] Strife, though, was not to linger far behind, and he soon fell out of favor with Schmidt. Subsequently sent to Germany, the ISB had been dissolved and a sizeable portion of Csallner's work destroyed by the time he returned to Transylvania in 1943.

131 Csallner, *Meine wissenschaftlichen Arbeiten* [1975], 7.

CHAPTER IV

FASCIST VISIONS OF A EUGENIC FORTRESS: THE SELF-HELP'S ORIGINS AND RISE TO POWER, 1922–33

Fritz Fabritius was to leave Saxon historiography with one of its most ideologically complex and polemically convoluted legacies by the time of his death aged seventy-four, on October 20, 1957, in Prien, near Hamburg. If Heinrich Siegmund had stood at the forefront of a budding eugenic discourse, and Alfred Csallner characterized its maturation into a political ideology in waiting, then Fritz Fabritius ultimately defined and drove a fascist movement for renewal that came to offer Saxon visions of a eugenic fortress a revolutionary alternative to the church—the Self-Help.

The astonishing metamorphosis of what began as a "workers' book club" in 1922 into the dominant force in Saxon politics by 1933, its ascension to the fore of the *Volksgemeinschaft* and hence leadership of all German minorities in Romania in 1935, the three years of "civil war" between the movement's moderates and radicals prior to their 1938 Berlin-mediated reconciliation, and ultimately the movement's *Gleichschaltung* at the hands of Andreas Schmidt in 1940 is a truly enthralling history. No matter how many times it was retold in Saxon historiography, this history rarely had its ideological core, its protagonists and antagonists, hauled to the fore of the debate. Strangely enough, the sorely needed biographies of many of the fascist movement's leading figures discussed here—including Fritz Fabritius and Wilhelm Schunn—are either unsatisfying or wholly absent, as are analyses of the movement's fundamental ideological character traits. Perhaps most regrettably of all, it is rather difficult to find any meaningful discussion of the policies (population or otherwise) that the Self-Help so consistently tried to translate into reality. Symptomatically, historiography is currently also waiting for an even elementary "structural" analysis of the Self-Help's membership and of the apparatus that, once empowered, it built on the foundations of existing political systems. Instead, the bulk of available literature tends to revolve around the omnipresent *Gretchenfrage* of the Self-Help's relationship with the church hierarchy, conservative circles, and prominent public figures (most commonly Rudolf Brandsch and Hans Otto Roth), or the matter of how, why, and when it enlisted in the ranks of German National Socialism (and the extent to which this signifies the adoption of ideology rather than the crystallization of an indigenous permutation of an international sea change).

This study will have to contend itself with offering a more constrained assessment of the Self-Help's rise and fall between 1922 and 1940. A far more detailed investigation of how its ideological makeup unveils a nascent fascist discourse emerging in harmony with the "new consensus" on generic fascism and its wider relevance as a crucial case study of an ethnic minority's quest to reinvent an alternative modernity to its particular needs is currently sorely needed. While it is simply impossible to embark on such an undertaking at this juncture, the investigation offered here aims to analyze the Self-Help's ideological and pragmatic quest to reinvent the Saxon nation, to smithy a "new," 800,000-people-strong nation out of the wide assortment of Germans to have become Romanian citizens after the First World War. In other words, the goal is to investigate the "movement for renewal" in terms of its ideological content, applicable means, and the reality of translating policy into practice.

1. Fritz Fabritius and the Origins of Saxon Fascism

Born on March 27, 1883, in Hermannstadt, Fritz (or, rather, Friedrich) Fabritius was to lay the ideological and methodological foundations for a nascent fascist Self-Help movement that repeatedly reinvented itself both structurally and ideologically during its brief but tumultuous existence between 1922 and 1940. So, who was Fritz Fabritius, how and why did he become the founder and figurehead of an indigenous Saxon fascist movement that went from book club to empowered regime within a decade? How has history remembered him, and to what extent is the prevalent understanding of his role in a nascent fascist—not National Socialist—bid to redefine and re-entrench Saxon cultural identity and Saxon racial anthropology in need of revision?

Fritz Fabritius's life and work was largely defined by the "Führer myth" he so fervently fed,[1] and that in the hands of his closest allies manifested itself as a celebration of his hands-on pragmatism—in contrast to the comparatively higher degree of ideological sophistication found with some of his contemporaries, such as Wilhelm Schunn, who amongst others authored the movement's 1931 twenty-five-point program, as well as the 1933 and 1935 national programs. The romanticized "rootedness" attributed to Fabritius contained far more political capital, and unavoidably lent itself to the customary "mythicization" of any revolutionary ideology's leading figure. So, for example, Otto Fritz Jickeli is found insisting "Fabritius was not a pondering theoretician, but leapt into his work with both feet. The new *Weltanschauung*'s fundamentals, which he had largely lived by already, were dogma enough for him."[2] And indeed, Fabritius, like so many of his peers locally and internationally, was not an ideologue but a disgruntled First World War veteran in search of new meaning.

Reflecting upon the Great War's impact on the "Saxon condition" pre- and post-1914, one of the most radical defamations of economic (and hence necessarily spiritual) degeneration came in the form of a 1928 theatre review in which Fabritius raved that "the World War was the deserved consequence for foreign [*Artfremde*] influence and the decay of our national life. The German spirit, that wants and needs to soar if it is not to die, can never ever want to be measured against the snatching [*raffenden*], rotting, egoistic spirit of the muddy Semitic worldview—it would, over short or long, be ruined by its contradictions of nature [*Naturwidrigkeiten*]."[3] Shortly thereafter, unsurprisingly, Fabritius published an appeal to all war veterans to join his cause.[4]

1 Fabritius's perception of what defined a *Führer* is relatively elementary when he writes that "the *Führer* must, if he wants to be a true leader, stand miles above the nation, be genuinely German in his simplicity, plainness, and sternness—only then will he last, only then can he work constructively on the national whole, only then will he demand a following's willing obedience stemming from conviction." Fabritius, "Führer," 1. Also, Fabritius clearly demanded complete submission when, two years earlier, he insisted: "You and your nation can only survive and thrive [...] if the individual willingly and believingly follows his leader, when the national community becomes a blood community again, whose discipline rejects everything spiritually and bodily foreign, when through racial purity [*Artreinheit*] the national unity [*Volkseinheit*] is rebuilt." Fabritius, "Tue ein jeder nichts als seine Pflicht," 1.

2 Jickeli, *Unser Weg zur Erneuerung*, 10.

3 Fabritius, "Deutsches Schausspiel," 2. In the same issue we also find an article applauding the Saxons' growing awareness of race-hygienic principles (as advocated by F. K. Günther), and insisting they should be used to assess the impact of immigration on the nation's biological organism. See "Die Verköterung der Menschheit."

4 Fabritius, "Frontkämpfer heraus!"

A graduate of the Viennese Military Academy, fluent in French and Hungarian, Fabritius rapidly moved through the ranks, making that of lieutenant in 1903. Nonetheless, he transferred to the reserves and moved back to Hermannstadt in 1907, the year he married Hermine Promper. In August 1914, Fabritius returned to active duty in a lancers regiment, before being transferred to the dragoons, was awarded the *Militärverdienstmedaille am Bande des Militärverdienstkreuzes* two months later, promoted to the rank of cavalry captain (*Rittmeister*) in March 1915, awarded the Order of the Iron Cross Third Class in October 1916, and had "swords" added to his *Militärverdienstkreuz* in February 1917.[5]

Despite this promising military career, the fascist myth-building exercise was never to focus on his wartime exploits. On the contrary, Fabritius was most commonly idealized for his trials and tribulations as a farmer after 1907, and for the "historic significance" of his 1912 anti-Semitic Youth Defense (Jugendwehr)[6] organization that purportedly had laid the foundations upon which the Self-Help would rise a decade later. Indeed, the perception of Fabritius as the broad-shouldered, compassionate, grounded, mild-mannered but driven *Führer* suffuses the following account offered by Wilhelm Schunn:

> Our movement's rooted history reaches far back. Because it begins as early as 1907. In that year, our current national leader Fritz Fabritius, aged twenty-four and a lieutenant, took off the emperor's coat and returned home to Hermannstadt. Those who knew him then know he was a beaming apparition that generally stood out in his distinguished regiment's uniform.
>
> That he left is no surprise, despite his love of everything military, because the army couldn't offer him one important thing: The possibility of influencing [*gesinnungsmässigen Einwirkung*] beliefs between two people. Apart from that, he yearned for farm work.[7]

---

5 NA Sibiu, CGR, A.I.1. Album A, p. 2.

6 It is clearly a reflection of the ideological climate when we find Schunn recounting that "Fabritius taught them that from now on Germanness [*Deutschtum*] must be their lives' greatest ambition. Jewish thought was discussed in depth and strictly denied. All of the boys became anti-Semites." Schunn, *Weg und Feinde der NEDR*, 9.

7 Ibid., 7. A more poignant, politicized version is offered by Otto Fritz Jickeli, who explained that Fabritius, "in realizing the value of blood and soil, took off his officer's cloak in 1906 [sic!] in order to become a simple peasant." Jickeli, *Unser Weg zur Erneuerung*, 67.

Upon his return to Hermannstadt, Fabritius had leased twenty-eight yokes of land in Am Salzburger Berg, which after two successive draughts had left him in financial crisis but, as Schunn continued:

> Even in his misery, and without intending to, Fabritius became the friendly spirit [*gute Geist*] of a crowd of simple, indeed of the poorest people that so naturally flocked around him—like a magnet attracts iron—and which he would have had to chase away had he not wanted to offer them anything. The mysteriousness of his personality lay with, as we later tended to realize laughing: One didn't like this, another that, and yet he is the point around which everything crystallizes, gathers, and without whom the Self-Help would have never emerged and thrived.[8]

This persistent emphasis on Fabritius's "genuine character" is omnipresent in the Self-Help's propaganda, and perpetually reinforced by the movement's leading figures, such as Transylvania's future *Gauleiter* Jickeli when he explains: "The magic of Fabritius's personality lies with his kindness that wins him every heart. His naturalness has always gained him the peasants' and workers' trust, for which he always was and will be 'our Fritz.' He has always lived with his coworkers like with a large family."[9] Another example of this romanticized portrayal of Fabritius as a truly humble and genuine soldier in the cause of Saxon renewal is offered by Helmut Wolff, who insisted "the nation immediately feels that here a man with a mission has entered their midst. A man who inherently rejects exaggerated intellectualism, who only judges people by their character, who knows of nothing other than to be a comrade amongst comrades, and to place all his labor in the service of the nation."[10]

In the propagandistic discourse, Fabritius's martyrdom ended with his 1912 encounter with the prominent banker Karl Wolff, a relationship that laid the programmatic groundwork for his future movement's form and function, and one of its most potent ideological founding myths. Following their initial meeting, the moment when Karl Wolff "realized who actually stood before him,"[11] Wolff gave Fabritius a job as a clerk in his Hermannstädter Allgemeine Sparkassa, where Fabritius not only gained particularly valuable

8 Schunn, *Weg und Feinde der NEDR*, 7–8.
9 Jickeli, *Unser Weg zur Erneuerung*, 69–70.
10 Wolff, *Ein Jahr Volksgemeinschaft*, 11.
11 Schunn, *Weg und Feinde der NEDR*, 8.

insights into the Saxon banking system, but more importantly, found a mentor and patron—one whom Fabritius was later to credit as the "father of the Self-Help thought."[12] Indeed, "the suffering was over. But Fabritius had to have been asked to put up with something terrible. The officer and enthusiastic peasant had to sit behind a desk without being able to move, and had to handle a frail quill in his mighty fingers, and every now and then at least pretend he knew what to do with it."[13] More pragmatically, Fabritius's ten-year employment with the bank unto his dismissal (or resignation, depending on the source)[14] in 1922, shortly after the Self-Help's creation, entered his public biography as a period of eager anticipation—a time when social upheavals reverberated through Saxon society in the wake of the First World War, when the desperate and disgruntled found themselves "magnetically drawn" to Fabritius, always their natural leader, if in the guise of a bank clerk.

Fabritius's appropriation of Karl Wolff as his *völkisch* father figure, and the Saxon periodicals' increasingly gloomy accounts of a postwar Germany in the grip of growing economic despair and spiritual decline created the backdrop to the Self-Help's second central (and historiography's most controversial) founding myth—the widely publicized chance encounter between Fabritius and Hitler in Munich in 1922. While this study (like so many others) has been quite unsuccessful in finding concrete evidence that such a meeting actually transpired, the fiction is itself in many ways more significant than reality. That Fabritius had been carefully tending to contacts in Austria and Germany before and after 1922 is widely known,[15] and that his choice in comrades leaves little to the imagination is similarly evident in his correspondence (if nothing else) with Guido List in February 1918,[16] his membership in a host of German organizations such as the Deutschbund, which he joined in March that year,[17] and his 1921 membership (interestingly enough as member number sixteen) in the Schirmherrschaft der Deutschen Bauernhoch-

12 Fabritius, "[Obituary] Dr. Carl Wolff," 1.

13 Schunn, *Weg und Feinde der NEDR*, 8.

14 According to Schunn, for example, Fabritius had been offered a choice and decided in favor of the Self-Help. See Schunn, *Weg und Feinde der NEDR*, 15. Jickeli claimed he had been fired without notice. See Jickeli, *Unser Weg zur Erneuerung*, 13.

15 What, if any, contact Fabritius had with German eugenicists is currently impossible to tell, but in 1931 he published a rather interesting appeal: "I need pictures of national brothers in our settlement area for a scientific study published [*herausgegeben*] by the Nordic Ring (Berlin). Send me two good passport photos. On the back side write name, city, date of birth, height, and skin, hair, and eye color." Fabritius, "Wer hilft mit!?"

16 NA Sibiu, CGR, A.I.1. Album A, p. 11.

17 Ibid., 12–13.

schule ev.[18] Similarly, there is evidence that Fabritius was exchanging letters (or, rather, postcards) with Ernst Röhm by 1926,[19] a time during which Fabritius also became increasingly involved with Wilhelm Kotzde-Kottenrodt and his Adler und Falken, a group which the Self-Help had hosted as guests earlier that year.[20] In any case, though, there is no tangible evidence of a correspondence, let alone a meeting between Fabritius and Hitler prior to May 1933, when Hitler thanked Fabritius for two unnamed books sent to him on April 13, 1933, and unremarkably concluded with "I am glad to read your letter's renewed insistence on the old unbreakable German *Kampfgemeinschaft* of the Germans and their ancestral homeland."[21]

Regardless, the various and varying accounts of how Fabritius met Hitler in Munich are of marginal value to an analysis of the Self-Help's early ideological development (no mention of them can be found anywhere in the *Selbsthilfe*'s issues published in the 1920s), while they were essential tools towards the movement's accumulation of political capital and ideological legitimacy after 1930, and can (currently) be best described as an attempt to backdate the movement's National Socialist credentials. But they, of course, also played a crucial role in the fabrication of Fabritius's own *Führer* myth, a process particularly evident in the accounts offered by Schunn and Jickeli, which hence merit a closer look:

> By this time, the Self-Help was already being influenced by National Socialism. The aged Saxon leader Dr. Karl Wolff raked some money together and sent Fabritius to Germany with the strange task of checking whether Germany had really sunken as low as the newspapers led one to conclude, and whether there was actually nobody left who wanted to believe in the German nation's future. During his search, Fabritius also found his way to Munich and into the National Socialist head offices. While debating with Aman, an unknown man joined, with whom Fabritius entered into an animated discussion. It emerged later that this man had been Adolf Hitler.
> Ever since this trip to Germany, Fabritius, and hence all his followers, were National Socialists.[22]

18 Ibid., 21.
19 Ibid., 28.
20 Ibid., 35.
21 Ibid., 48.
22 Schunn, *Weg und Feinde der NEDR*, 13–14.

More polemically saturated still, Jickeli argued that destiny itself had moved Karl Wolff to send Fabritius to Germany, where he again unwittingly spent hours musing with Hitler in the *Völkische Beobachter*'s editorial offices:

> It happened in the *Völkische Beobachter*'s offices, when Fabritius met a young man with whom—sitting on a wooden crate—he talked for over an hour, and who completely drew him into his thoughts' spell. He had not understood his name when introduced. It was only on the next day, when reading a magazine, that he realized he had met Adolf Hitler, who was yet at the beginning of his struggle.
>
> And so Fabritius was probably the first *Auslandsdeutsche* to have unequivocally joined the National Socialist ideology [*Weltanschauung*].[23]

This insistence on National Socialist origins is, as such, a fabrication, a distortion of the indigenous ideological roots of the Self-Help that while feeding off essentially the same conceptual nutrients, grew along a different tangent to its German counterpart. That was the case prior to the movement's division into economic and political wings ten years later, in 1932, when the Self-Help did indeed wholly appropriate the NSDAP's franchise by rebranding itself as the National Socialist Self-Help Movement of the Germans in Romania, while its mouthpiece *Selbsthilfe* was renamed the *Osdeutscher Beobachter*. But when the Self-Help published its first flier in June 1922, its aims were far humbler: "A home on one's own soil, with air and light, with fruit and flowers and sunshine and economic freedom and independence without exploitation—that's the path we see towards our goals."[24]

II. Early Development, 1922–29

So how and why did the Self-Help evolve so rapidly from a workers' book club grounded in Heinrich Dolle's principles of individual economic autarky expounded in *Aus Not zu Brot*[25] into a fascist ideology saturated with eugenic themes so common to the work of Siegmund and Csallner, an ideology that within a year of forming a political party found itself at the helm of Saxon

23 Jickeli, *Unser Weg zur Erneuerung*, 65.

24 Dolle, "Unser Ziel."

25 Although Dolle vanishes from the Self-Help's discourse rather quickly, his book's importance to early ideology is also acknowledged by Schunn. See Schunn, *Weg und Feinde der NEDR*, 11–12.

politics in November 1933? In short, the Self-Help underwent three distinct phases between 1922 and 1940. The first, 1922–29, was one of relative obscurity and moderate growth, but characterized by the ongoing refinement of an increasingly political ideology geared towards attracting the urban workers; ever more apocalyptic visions of national degeneration and disenchantment with the status quo; and a growing emphasis on eugenic population policies. During the second phase, spanning from 1929 to 1932, the movement experienced a profound diversification of its membership, when a host of existing societies and organizations joined its ranks (such as Csallner's VdK and Alfred Bonfert's Wandervogel), accelerating its ideological radicalization and setting the stage for the movement's explicit adoption of National Socialism in 1932. The third phase, 1932–40, consequently began with the Self-Help's partition into an economic and a political wing following polarized internal debates. Although the decision to leave the lobby trail in favor of party politics was unpopular with the likes of Wilhelm Schunn, the newly created NSDR was to expand its power base and influence with such speed that a mere year later it rose to political domination. In October, the fifth and last Sachsentag adopted its compromised national program, the elections for Saxon national bodies a month later handed them the Saxon National Council, and Fabritius was elected chair of the Association of Germans in Romania in June 1935.

The Self-Help thus began as an urban movement that like so many of its peers looked to the peasant as a source of regeneration, while initially focusing its attention on attracting the urban poor, and insisting in later years that it had been the first to locate and organize a comparatively new but growing Saxon social estate: the workers. In April 1922, the Self-Help first entered the public domain with a flier announcing the creation of the Saxon Workers' Book Club, founded to lead the nation towards "true Saxon nationalist [*völkisch*] socialism," and calling for donations towards the publication of a newsletter entitled *Freedom and Brotherliness*.[26] The 700 lei that the reading group collected, though, were put towards the publication of the movement's mouthpiece, the *Self-Help: Kampfblatt for the Honestly Working Nation*.[27]

26 Glass, *Zerbrochene Nachbarschaft*, 323–24.

27 The Self-Help's mouthpiece, the *Selbsthilfe: Kampfblatt für das ehrlich arbeitende Volk*, underwent numerous illustrative modifications during its twelve-year print run between 1922 and 1934 (though it remained with Hermannstadt's Botchner Verlag throughout). Resembling a disjointed journal at first, the *Selbsthilfe* published a mere thirteen issues between June 1922 and December 1928. Reflecting the movement's rapid ex-

The blueprint of an embryonic fascist movement that came to define itself, first, in terms of a dual palingenetic quest to reinvent and purge the Saxon nation, and second, in terms of the considerably more utopian nation-building project of uniting all of Romania's Germans into one 800,000-people-strong nation, is clearly visible in the Self-Help's first organizational incarnation, the rather ungainly 1922 Union of Small Farm Breeders, Gardeners, Land, and Homestead Hungry (Die Vereinigten Kleinviehzüchter, Kleingärtner, Land- und Heimstättenhungrigen). As with Csallner, the degree of food production by city dwellers was a crucial ideological imperative to a movement that strove towards self-sustainability. But while this heavy emphasis on encouraging private attempts at small-scale subsistence farming and breeding small farm animals is typical of Dolle's approach to tackling poverty, the Self-Help's mission statement foreshadows a considerably more political ideology when explaining its aims thus:

> What it wants is hinted at with our title. Our economy has been destroyed. Thoroughly! Let us now build our own land of children [*Kinderland*]. We continue to live in our children, as our parents live through us.
> Destroyed is also our faith that anyone, from anywhere, from parliaments, parties, leagues of nations or their like, anyone from the outside, could bring us salvation and healing, or that salvation and healing could come from any of the previously trodden paths. We need to create it through and out of ourselves.[28]

That the Self-Help was always intended as a proactive, political organization (if not a party), above and beyond decrying degeneration, is illustrated by an appeal published in the same *Selbsthilfe* issue, and the first to include biopolitical terminology in its definition of the regenerative agenda: "We will help ourselves. We will create healthy germ-cells for a healthy nation, and will elevate this nation above the station of the decaying state full of parasites, will replace the top-down rule with the natural and healthy growth from the bottom up."[29] Although the Self-Help, unto

pansion during the late 1920s, it subsequently appeared on a monthly basis, before becoming a biweekly paper in October 1930, and a weekly in June 1931. The singularly most significant change, though, came a year later, in June 1932, when the *Selbsthilfe* became the *Ostdeutscher Beobachter* (but kept the subtitle), reflecting the movement's rebranding as NSDR earlier that year. Although the paper also witnessed the NSDR's transformation into the NEDR, it ceased publication after the "movement for renewal" was banned in July 1933.

28 "An die deutsch-sächsische Bevölkerung."

29 Dolle, "Wie soll diese Selbsthilfe aussehen."

1929, frequently appears to have offered little more than an umbrella under which various individual suborganizations and agendas aggregated (rather than constituting a single streamlined movement), it actually subsumed three individual components by July 1922: the original Saxon Workers' Book Club (Deutsch-sächsische Arbeiter-Lesezirkel), the Settler Group (Siedlungsgruppe), and the Economic Group (Wirschaftsgruppe).[30] This internal division of labor points to a fundamentally important, if commonly overlooked triad of in(ter)dependently pursued goals: the movement's early focus and attempts to recruit members from a growing urban class of workers; internal colonization work and the centrality of the *Lebensraum* motif; and the omnipresent condemnations of capitalism and the international banking systems.[31]

If only a handful of disjointed *Selbsthilfe* issues were published in 1922, and its appearance was suspended due to a police ban in 1923 and 1924,[32] 1925 was to not only mark the movement's return to the public stage, but the journal was now espousing a markedly radicalized discourse.[33] The year's first issue began with a remarkable vision of the Saxon nation's accelerating descent into oblivion:

> And what does the Self-Help want? These are once more difficult days, and all sorts of hardships are pressing in. The good old tradition has been forgotten, misunderstood "new-ageism" [*Neuzeitlichkeit*] has confused the mind, and an all-encompassing, rampant materialism has poisoned the hearts. An unnatural *Genussleben* spreads everywhere and threatens to swamp and suffocate a healthy national life. [. . .] The Self-Help must grow [. . .] until the small community has become a real nation of brothers. Then we will once more have spiritual and moral bulwarks in healthy people with a view of life against which the muddy waves of a destructive time will lash in vain.[34]

30 "Volksgenossen," 4.

31 A valuable early example of these indictments is offered by the article "The New Socialism!" rabidly attacking internationalism, communism, and the Saxon political elite as failed systems. See "Der Neue Sozialismus! Teil I."

32 Glass, *Zerbrochene Nachbarschaft*, 323–24.

33 1925 also sees the journal's first contribution to specifically discuss the question of "Race and Nation," in which H. Greisling delights at how the "racial question" had finally begun to spill over from Germany on the back of a growing *völkisch* movement, which would help revive a "self-preservation instinct" undermined by the "new-age decadence." Greisling, "Rasse und Volkstum."

34 E. C., "'Selbsthilfe.'"

This professed quest to redefine, regenerate, and reground the Saxon nation might as well have been taken from a speech in the mid-1930s, and clearly illustrates that the movement's ideological foundations had been laid long before any of its discourses so much as mention National Socialism. That said, what tangible outcomes did the Self-Help achieve during its first phase, between 1922 and 1929, above and beyond recruitment campaigns and the publication of fliers? In addition to the activities of its economic and colonization sections discussed below, by 1926 the Self-Help had begun implementing an internal population policy agenda, be it in the form of largely symbolic "Christmas gifts" of 500 lei awarded to the four largest families, or through the continuous organization of lecture series and training courses geared towards the "territorial expansion absolutely necessary for life—the strengthening of the spiritual and racial spheres [*Belange*]."[35]

According to its own figures, the Self-Help founded three settlements in Hermannstadt by August 1925:[36]

1. Demargrund had been acquired as early as October 1922, and had twenty-six plots, ten of which had completed houses, with four more under construction, in addition to a communal well. By the end of 1926, under the guidance of the settlement's director Viktor Gödri this number increased to eighteen completed and eight "half-finished" homes.
2. Heltauerstrasse, with thirty-two plots, had been bought in May 1924, and had two completed homes with a further three under construction; by December 1926 only six homes had been built.
3. Schellenbergerstrasse, the largest of the three settlements, spanning fifty-three building plots, was purchased in June 1925 but had no homes under construction in 1925 or 1926.

These colonization projects hence predate the Self-Help's Building Society (Baufond), founded in June 1925, which by December that year had

35 Fabritius, "Die Arbeit die unsere Vereinigung leistet."

36 "Bisher geleistete Arbeit." See also Fabritius, "Welche Arbeiten hat die Selbsthilfe bisher geleistet?" The data for 1926 is taken from Fabritius, "Die Arbeit die unsere Vereinigung leistet." See also Schunn, *Weg und Feinde der NEDR*, 13. Also see the report on the meeting of the Self-Help's Hermannstadt Regional Branch on October 25, 1929, found in ZAEV, LK 103.90 (1929), 19.

already distributed 100,000 lei in interest-free loans.[37] While the numbers do not account for the number or size of the individual loans, it appears safe to conclude that their clientele was rather small, given that by the end of 1926 the fund had attracted only sixty-one members, the year in which a total of 635,000 lei were handed out in loans to eight Self-Help members.[38] By 1927, the total sum lent in the two-year period between October 13, 1925, and September 23, 1927, rose to 1,530,000 lei,[39] growing to 2.5 million lei by the end of August 1928.[40]

Complementing this exponential growth in reported outcomes, a considerably clearer, conceptually richer discourse was crystallizing by 1927. Saturated with the themes and visions common to the work of both Siegmund and Csallner, it emerged from the pages of the *Selbsthilfe*. The growing demands for ethnic homogeneity and an expansion of Saxon *Lebensraum*[41] are illustrated particularly clearly in Fritz Fabritius's article "Homeland":

> Which are the foundations that guarantee any national community's continued existence? Apart from the community's homogeneity (racial homogeneity [*Rassegleichheit*])—which is the natural precondition for shared moral values [*sittliches Empfinden*]—a shared sense of law and honor, and its communal economic policy, the availability of sufficient *Lebensraum* is the central issue from which all paths towards a healthy higher evolution [*Höherentwicklung*] lead.
> The Self-Help is founded upon these preconditions. It wants, in the first instance, to nourish and strengthen the national body's germ-cell [*Keimzelle*]—the family—economically and nationally [*völkisch*] in order to begin strengthening the entire national community through this painstaking work.

37 For the fifteen-paragraph-long guidelines for Self-Help loans from the Hermannstadt Business and Merchant Bank (Hermannstädter Gewerbe- und Handelsbank AG) see Fabritius, "Baugenossenschaft: Teil I" and "Baugenossenschaft: Teil II." The creation of the Building Society was, at least according Schunn, a stroke of genius, in that it allowed for the provision loans without having any starter capital: "This simple and selfless group had, with the form of a building society but without any cash and previous models, found Columbus's egg." Schunn, *Weg und Feinde der NEDR*, 15.

38 Fabritius, "Die Arbeit die unsere Vereinigung leistet," 1–3. Of these ten loans: one for 20,000 and one for 30,000 lei; two each for 40,000 and 50,000 lei; one to the sum of 80,000 lei; two of 100,000 lei; and a final one for 125,000 lei.

39 See "Verschiedenes."

40 See "Liste der ausbezahlten Beträge."

41 For example, Erna Shaper's "Nation, Make Room for Yourself," in which she demands the reader: "Make sure that the holy land of your fathers is really yours! You will owe your grandchildren an explanation! [...] If someone needs to retreat, so let non-Germans be the first to leave." Shaper, "Volk schaff dir Raum," 3.

> National regeneration [*Ertüchtigung*], raising it to selflessly serve its own blood community [*Blutsgemeinschaft*], and expanding our settlement areas must therefore be the central objective [*das Um und Auf*] of any political or economic policy.[42]

The year 1927, in which Siegmund joined, also saw the Self-Help's first attempts at formulating and advancing policy proposals, and the first of these argued for the creation of "settler schools" (*Siedlungsschulen*) by the Saxon National Council (but, indicatively, managed from a central office in Germany) in all German settlement areas. In effect, these schools would have selected six to twenty apprentices (half girls, half boys) from "probably poor, preferably large, industrious, and racially flawless families." These were then to be housed and educated for six years, funded from the public purse (i.e., the church's youth organizations), for which they in turn offerrd their labor for free before being temporarily sent abroad on an exchange program. Fabritius ultimately hoped such schools would both assist "worthy" families and allow for "the mass of these thus-educated master craftsmen to return home and to be planted into their place of birth or a hard-pressed area as the seed of a new movement for renewal."[43]

In 1928, the Self-Help transformed itself into a "legal body" by founding the Self-Help Company for the Mediation and Financing of Real Estate Fr. Fabritius & Co. Limited Partnership in Hermannstadt (Selbsthilfe, Unternehmung zur Vermittlung und Finanzierung von Immobilien Fr. Fabritius & Co, Kommanditgesellschaft in Hermannstadt).[44] The company's charter offered perhaps the closest thing to a movement program, and opened with the Self-Help's three guiding economic principles: "man always stands miles above money," "money creates no values," and "it is unnatural for our German economy to impose interest."[45] On the other hand, the company's underlying beliefs are subsumed under two key points: first, that "every honest German has, in our eyes, the same value for our national community if he is straight [*geradlinig*] of mind and body.

42 Fabritius, "Heimaterde," 1.

43 Fabritius, "Siedlung und Siedlerschulen," 1–2.

44 For the twenty-paragraph-long guidelines and procedures outlining its organization and payment schemes, see "Richtlinien der 'Selbsthilfe'"; and "Geschäftsordnung der Firma 'Selbsthilfe.'"

45 Fabritius, "Selbsthilfe," 2–3.

Rank, standing, and certainly wealth are secondary so long as we are dealing with a man healthy in mind and body"; and second, highlighting the dogmatic motif underlying its *raison d'être*, "no one other than the national community should benefit from our work, which is why we neither give nor take interest."[46]

III. Expansion and Radicalization, 1929–32

It is during 1929 and its aftermath that the Self-Help undergoes the fundamental changes that paved the way to becoming an overtly National Socialist political party in 1932. The first and perhaps most important point, given the secondary literature, is that only in 1929 does the Self-Help begin to publish an increasing number of articles borrowed from non-Saxon periodicals,[47] amongst which one finds a flurry of contributions taken from the *Völkischer Beobachter*.[48] By 1929, the Self-Help was also growing ever faster, with roughly 400 members representing ten local branches congregating for the movement's "extraordinary general assembly" on February 24, 1929, in Hermannstadt.[49]

The degree to which the movement's expansion had been driven by its ideological-political, rather than economic appeal to the wider populous is apparent in the emphatic report on the activities of the six-month-old local branch in Kronstadt, published later that year, in which its leader (*Obmann*), the pediatrician Gustav Waber, proclaims: "Today the Self-Help is being discussed everywhere—whether in happy agreement or derogatory and dismissive, is largely irrelevant! Because the majority needs the Self-Help and flock to it—unstoppably! And the leaders—they who to this day try to skirt around this incredibly important question with failed measures—realize in horror that once more the masses are about to slip through their fingers, that the Self-Help is marching inexorably and incessantly, is gaining ground, Saxon ground!"[50]

46 Ibid.

47 Including, amongst others, the *Evangelisches Deutschland*; the Swedish *Nationen*; *Schweizerbanner*; and the *Schwäbische Landjugend*.

48 The very first of these articles was "Der anwachsende Bankrott der deutschen Republik." In 1929, the journal also published a number of race-hygienic articles borrowed from the *Völkische Beobachter*'s pages, three of the most interesting choices being: "Was der Laie von der Erbforschung wissen sollte"; "Die Nürnberger Tage"; and Oberdörfer, "Bluterneuerung der arischen Rasse."

49 See "Ausserordentliche Generalversammlung."

50 Waber, "Ortsgruppe Kronstadt: Teil I," 5.

Waber continued to recount how the branch had been founded by the carpet trader Michael Orendi, who upon hearing of the Self-Help had travelled to Hermannstadt to consult with Fabritius, and during their conversation invited and paid for Fabritius to come to Kronstadt and deliver a lecture on December 17, 1928. The local branch was founded ten days later, and grew to 100 members in the following six months. This account not only reinforces Fabritius's central role to the movement's ability to attract members, but also reflects the Self-Help's ideological expansion above and beyond purely economic pursuits when discussing the branch's quarterly "scientific" lectures. Interestingly enough, following Fabritius's initial speech, the second, in May 1929, was Alfred Csallner's "The Rise and Fall of Nations," outlining his biopolitical work and underlining the Self-Help's significance as a moral and economic tool of "national defense" (*Volksverteidigunsmittel*).[51]

The following year, in 1930, Waber published a particularly telling article entitled "Thoughts on Racial Purity," decrying the degenerative impact of mixed marriages—in fact, the first contribution to the *Selbsthilfe* to do so. After lamenting the loss of human capital that mixed marriages imply, Waber makes the following appeal to work together on the road towards race-hygienic regeneration and exclusivity:

> Comrades of both sexes, peasants and townsfolk, defend yourselves against the mixed marriages that are getting out of hand, because they spoil our German nature [*Art*], reduce and devalue our offspring, whose hundreds-year-old seeds of industriousness are damaged. Much rather, help—everyone in their own way—to encourage marriages between the same-blooded [*Gleichblütigen*] and, further, to remove any supposed obstacles on our newly and successfully embarked-upon path towards large families [*Kinderreichtum*]! Know this: the Self-Help is happy to work with all of you towards the realization of these race-hygienic demands. Close the circle![52]

Apart from the apparent absence of economic policies in any of these speeches, the Self-Help's membership numbers similarly testify to the extent to which the movement was not only inherently political, but that the political exerted more of a pull than its economic policies: of the 1,077 new

51 Waber, "Ortsgruppe Kronstadt: Teil II," 6.
52 Waber, "Gedanken zur Reinhaltung der Rasse," 2–3.

members gained in 1930, only 498 also joined the Building Society, which at this point only accounted for roughly half of the movement's active adherents, namely 1,448 out of a total of 2,697.[53]

The Self-Help's rapid expansion in the late 1920s and early 1930s was significantly bolstered by various organizations entering its rank and file, most importantly here Alfred Csallner's Society of Child Enthusiasts in 1929, but also Alfred Bonfert's Wandervogel-Youth (Wandervogel-Jugend)[54] and Misch Bergleiter's Young Farmers (Jungbauern)[55] in early 1930 (all three of which subsequently published supplements in the *Selbsthilfe*). The extent to which these new recruits were to drive the Self-Help's ideological radicalization is fittingly foreshadowed by the perception of a nation in cataclysmic decline offered by Misch Bergleiter in 1929:

> It is getting ever darker. Like a poisonous fog it rises everywhere in the houses of the great, in the huts of the meagre, rotting and poisoning all old German customs and traditions. Freely and openly, with an impertinent smile impropriety crosses the street, and increasingly enslaves the individual of all classes. What the fathers have created through honest work, under the sons it is being suffocated by the strangling chains of an eerie power. And already treason spreads within the nation's ranks. All of this storms over the German nation, our nation, and upon so many descends the night of fatigue, despair, and hopelessness.[56]

Amongst the flurry of new recruitments, the incorporation of the veterinarian Dr. Alfred Bonfert's Southeast German Wandervogel was greeted particularly enthusiastically, as it gave the Self-Help a preestablished youth movement. One of Fabritius's early protégés, whom prior to the rift in 1935 Schunn considered "the educational ideal incarnate upon whose template our youth should be raised,"[57] Bonfert filled a vital gap in a movement that

53 Fabritius, "Kameraden!" See also Schunn, *Weg und Feinde der NEDR*, 18.

54 Fabritius, "Aufbauarbeit."

55 Fabritius, "Kameraden!"

56 Bergleiter, "Deutsche Weihnacht," 1. On a rather more poetic note, and as Csallner had so frequently done, Bergleiter summarized his perception of the nation's decline in the form of a poem ending with the verse: "Where are you, German nation? Are you standing in front of the abyss / that yawns wide before your feet? / Have you come to the end of your earthly life? / Are you worn out, racked nerves [*entnervt*], like a wilted leaf, / that every breeze can impertinently [*frech*] blow you away?" Bergleiter, "Vorspruch," 1–2.

57 Schunn, *Weg und Feinde der NEDR*, 48.

had recruited most of its leaders from Fabritius's First World War veteran generation, and its members from the urban poor. The Self-Help's expansion into youth work was, of course, seminal to a movement built upon a regenerative agenda that had moved on from urban farming practices, and was to place the advancement of "youth voluntary work" at the heart of its educational and social equality discourses.

The concept of "voluntary work" and its corresponding "work camps," according to Fabritius, had the underlying goal of "inserting an educational step between school and work life, in order to raise our comrades to think and act in a communal spirit, and in order to create new national wealth through voluntary work."[58] An illustrative summary of what the Self-Help largely understood the youth's education to subsume—and an apt description of the apparent symbolism featured on the 1926 postcard commemorating the Youth Movement Day in Heltau—is again provided by Misch Bergleiter, when he insists: "That which we must hammer and hammer again into our youths' mind is to believe [*Glaube*] in our nation's future and the German man's superiority. The heroic and the soulful in our peasants, which have been lost over the centuries, must be revived and, together with the sense of responsibility, [must] embark upon the path of national reconstruction [*Neubau*]."[59]

Schunn, who considered "voluntary youth work" to be "one of the most beautiful of gifts amongst the many beautiful things to have come to us from the motherland in these past years," continued to distil its wider educational and propagandistic purpose into the following key aims and ambitions: to accustom the individual to work for his community; to leave the respective community for which they worked with the knowledge that the nation is mobilizing, and thereby encourage them to work harder; to create material goods (mostly agricultural infrastructure) that otherwise had not been created; to provide an educational tool furthering the participants' "appropriately German" (*artgemäss deutschen*) way of thinking; and to offer the

58 Fabritius, "Arbeits-Dienstpflicht." The Self-Help was, of course, not the only fascist ideology to insist on the educational value of youth "work camps." Their comparison to the model advanced by the Romanian legionary movement would be particularly rewarding, and is likely to shed some much-needed light on the degree of ideological or even structural interaction between the two movements. While the detail such a study would require is prohibitive here, see Rebecca Hayne's valuable analysis of the legionaries' quest to hone a "new man" in and through its "work camps": Haynes, "Work Camps, Commerce, and the Education of the 'New Man.'"

59 S., "Gautagung in Hermannstadt: Teil II," 3–4.

unemployed a worthwhile task. But the three perhaps most important anticipated outcomes of spreading voluntary youth work were, first, to give "our nation's bourgeois [*verbürgerlichte*] face the hard features of a worker"; second, to bring about a "spiritual" balance and sense of belonging between rural and urban "sons and daughters of all classes through the comradery of communal life in the camps"; by, third, mediating between the two and furthering a sense of communal destiny and *Heimat*.[60]

Bonfert's Wandervogel had organized the first of these "voluntary youth work camps" in August 1931, in the village of Meschen, which was tasked with digging a drainage ditch.[61] Assessing its outcome, Schunn emphatically concluded: "And the result? A swampy area has become valuable meadows. City boys and girls have tans, are sinewy, and have returned home proud of their achievements with the resolve to raise their nation again. The student and worker have become friends for life. And the entire village's surroundings have become National Socialist."[62] Whether these targets had indeed been achieved is almost irrelevant compared to the underlying ideological value that they exhibit: to not only save and remold the nation's hereditary stock, but to regenerate and re-root its *Lebensraum*.

The following year, in the summer of 1932, the movement set up the Self-Help youth work teams (*Selbsthilfearbeitsmannschaften*, or SAM),[63] which were to become one of the movement's most important and simultaneously most embattled arms. Initially called the SA, the name was soon changed to SAM following German reservations about the obvious affinity to the NSDAP's SA—a comparison very much encouraged by the uniformed marches that the SAM so commonly staged at Self-Help rallies.

The Self-Help's rapid expansion in size and influence also attracted an increasingly vocal, if largely peripheral opposition from conservative circles. A characteristic example of the opposition's tendency to focus on the Self-Help's economic policies rather than its rapidly radicalizing ideology,[64]

60 Schunn, *Weg und Feinde der NEDR*, 26.

61 Its aims and ambitions are defined in Bonfert, "Das Arbeitslager in Meschendorf."

62 Schunn, *Weg und Feinde der NEDR*, 29.

63 See "Selbsthilfe-Arbeitsmannschaft." It also includes the SAM's Order no. 1, covering their form and function, uniforms, fiscal policy, etc.

64 In fact, the question should include the matter of whether, as some Self-Help members believed, this was a deliberate tactic to first silence, then to praise and sideline the movement into purely economic issues. See T., "Unser Programm."

a "damning" critique published in the *SDT* in July 1930 set out to dismantle the economic viability of signing up to their saving account scheme, accused of enticing poor farmers into living beyond their means. Symptomatically, Fabritius's reply, entitled "*Weltanschauung* Matters," insisted that while their loans do, after all, require a "modest" level of interest at repayment, they were nonetheless a far cheaper source of money than the high-street banks.[65] More importantly though, Fabritius was also to insist that regardless of whether repayment deadlines or minor interest rates could or would accrue, the Self-Help was effectively rendering the nation a public service that lifted the desperate poor out of their "basement squalor" and into "light and airy homes,"[66] hence a service connected to a wider agenda of social health and hygiene that, if understated, had been the movement's founding motif.[67]

This peculiar stance that the Saxon elites initially assumed towards the Self-Help is replicated across the board, where one frequently finds Fabritius's contemporaries, at least until 1931, attacking the Self-Help's business model while praising its national, *völkisch* influence and achievements. Looking on with what at times appears to be benign curiosity, one of the Self-Help's future archnemeses, the *Kronstädter Zeitung*, is found pondering in September 1930 that "one will have to discuss [*auseinandersetzen*] many of the details with Fabritius later. But it may remain uncontradicted that his movement has become an ethic fountain of strength for us, one whose simplicity just might create a true community [*Gemeinschaft*]."[68] Similarly, Hans Kaufmes is found lauding the Self-Help's creation as a "*völkisch* act! Who would have thought it possible to create a movement whose central aim is to encourage people to save," while Bernhard Capesius (who was to criticize Csallner's eugenic stance on inherited talent) ponders "the suggestive influence of the new [. . .] it would be difficult to imagine a greater con-

65 Fabritius, "Weltanschauungsfragen."

66 See also Wagner, "Ansprache." While Wagner is rather more apologetic, and in referring to Ludwig's report insists that the movement was still young and searching for the most appropriate mode of conducting its essentially Christian neighborly work, the dualism between economic means and ideological imperatives is omnipresent. So, for example, the subsequent issue featured an article linking, as always, the two themes, when it insists: "National Socialism is making tremendous advances; it seems to me to be the German human's reaction to the subhuman who only sees salvation in money and record profits." Dr. R. E. F., "Der Kampf der Geister."

67 A point forcefully made by Gauleiter Dr. Waedt when, addressing the Self-Help's regional assembly in November 1930, he derides the "ghastly hygienic conditions" in larger cities. See Schunn, "Die Gautagung in Schässburg."

68 "Die 'Kronstädter Zeitung,'" 3.

trast than that between the overflowing and spirited meeting of these young and ambitious [*aufstrebenden*] men, and the sparsely attended, dull session of the historical section of the *Landeskundeverein*—while that of the natural sciences was cancelled due to the lack of participants."[69]

Indeed, Fabritius repeatedly insisted that the Self-Help was not merely economic in nature. During a speech given to the Hermannstadt local branch, he proclaimed: "The Self-Help is not a business, but a homogenous community [*Artgemeinschaft*] in life and death, bearing the fervent will to pave a path leading our nation from peril and collapse towards freedom. [...] The Self-Help has become a 'religious-union' [*Glaubensgemeinschaft*] where every fighter is duty bound to stand up for his 'like-minded comrades' [*Gesinnugsgenossen*] with his entire being, even his life."[70] In fact, this period in the movement's conceptual evolution witnesses the first use of religious terminology to define the Self-Help's increasingly militant self-perception as a community driven by the mystical quest to salvage the nation's body and soul. A truly remarkable example of this is found in the *Mediascher Zeitung*'s report on the movement's regional assembly (*Gautagung*) in Schässburg, on November 2, 1930:

> Life has been brutalized and has lost its way. The Self-Help marches as a true [*rechter*] apostle through town and village. But help must come from villages where naturalness can still be found amongst the peasants. It will be hard work to integrate all Saxon farmers. Hail you, you who call!
> It was a celebration of awakening. A German sermon of unity and will and strength. A truly German missionary sermon in the Savior's spirit.[71]

It was also during this meeting in Schässburg that Csallner increasingly moved to the movement's fore, addressing the crowd with a speech deriding the growing threat Romanian expansion posed to Saxon *Lebensraum*, and pointing to the town of Braller as a key example of the destructive repercussions that the loss of ethnic homogeneity incurred.[72] Two weeks later, Csallner also featured in the program of the movement's two-day "regional assembly" in Hermannstadt, where in front of the estimated 2,500 attendees he

69 "Pressestimmen über die 'Selbsthilfe,'" 2–3.
70 "Mitgliederversammlung der O.G. Hermannstadt," 4–5.
71 See "Selbsthilfe," *Mediascher Zeitung*.
72 Schunn, "Die Gautagung in Schässburg," 4.

recounted how the Self-Help had been founded in 1922, how it had expanded into other German settlement areas in 1928, that Hermannstadt alone would now boast 1,105 members, that the Self-Help had already paid out over sixty million lei in loans, that it was at the front of the vital battle to regenerate the Saxon *Lebensraum*, and so on.[73] By the following summer, Csallner had risen to the post of *Gauleiter* Hermannstadt (with Dr. Wilhelm Hager still the leader of the movement's local branch), giving a detailed report of his travels and lectures throughout the region to the Hermannstadt regional assembly in June 1931.[74] Invoking the virtues of abstinence and saving, Csallner derided the employment of non-Saxons and decried the loss of an estimated eighty million lei that Saxons presumably spent annually on alcohol and tobacco—a sum he believed easily capable of covering all national and cultural expenses in addition to allowing for the repurchase of 35,000 joch "lost" to the agricultural reforms in the early 1920s. Pressing the issue further, Csallner continued to advertise the achievements of the "saver communities" he had fostered, and especially those of his local branch in Stolzenburg, claiming its twenty-seven members were saving 6,000 lei annually (with which they paid the bulk of their church taxes).[75]

Apart from the usual and customary "save the nation" rhetoric, perhaps the best illustration of the eugenic discourse's growing centrality to the Self-Help's ideology is offered by Fritz Fabritius in one of the long list of articles entitled "What We Want," in which he complains that:

> Foreign and unnatural ideologies, foreign law, and a foreign global economic crisis have diseased and hollowed the most valuable in our nation, our soul and sense of justice, and led us step by step on our destined path towards chaos, tumbling towards the abyss. [...] The ever-further-reaching racial degeneration and bastardization [*Verköterung und Verbastardisierung*] of the German nation has increasingly dwindled our great race values and blood values, so that now even the last sense of worthy, emotional deeds stemming from the blood's wants [*gefühls- weil blutgewollte Handeln*] has been lost.[76]

73 See "Veranstaltungen des Gaues Hermannstadt der Selbsthilfe"; and "Unser Vormarsch."

74 S., "Gautagung in Hermannstadt: Teil I."

75 S., "Gautagung in Hermannstadt: Teil II."

76 Fabritius, "Was wir wollen." Fabritius subsequently published his "Ten Commandments for Every Self-Help Man," which addressed the matter of racial hygiene as the eight of his ten rules: "Do not waste your racial inheritance [*Rasseerbgut*] on inferiors, because this is a sin against the spirit and a grave blood disgrace [*Blutschande*]." See Fabritius, "10 Gebote," 4.

That said, Fabritius continued to define his three fundamental demands: first, "to reawaken our nation's natural and blood-appropriate way of thinking, our organic sense of community"; second, that this pursuit of nation (re) building necessitated the embrace of struggle as a form of creative destruction, that the Self-Help was hence a *Kampfgemeinschaft* because "only struggle is creative, because only life able to assert itself is that which wants to fight for its place in the sun, because only he masters life who is willing to fight for an idea with his life"; and third, that "money once more becomes the servant of man, rather than man that of the fake-God [*Götze*] Gold."[77] If an explicitly economic agenda is to be found in these words, then it must be Fabritius's belief that charging interest on loans was foreign to the German spirit, a degenerative import to which the Saxon banking system had succumbed in the interwar period, and an unacceptable departure from Karl Wolff's economic principles of banking altruism.

Further to the insights into the Self-Help's radicalizing ideology, this article had an added significance in that it also appeared in Heinrich Zillich's seminal cultural journal *Klingsor*,[78] and marked the first meaningful literary exchange between these two groups (other than Zillich's initially reserved, while sympathetic views on the Self-Help), and foreshadowing their joint convening of the Meeting of Saxon Brothers of All Estates and Settlement Areas (Tagung sächsischer Volksgenossen aller Stände und Siedlungsgebiete) on June 7, 1931, in Hermannstadt.[79] At this meeting, Klingsor and the Self-Help concluded that the political status quo had comprehensively failed, and consequently demanded it convene a Saxon Day to discuss and decide upon a new national program.[80]

Clearly, 1931 was to become a cataclysmic year for the Self-Help, one in which speeches increasingly turned into marches, and one in which Fabritius insisted that the time had come to sink or swim: "If this year the good and constructive powers do not find the way to heal— that is, to eradicate all that is sick in our community through relentless battle—then we truly

77 Fabritius, "Was wir wollen," 1–2. Böhm labelled Zillich the "founding father [*Vordenker*]" of German fascism in Romania, and it is interesting to note that Zillich founded the Saxon postwar umbrella organization in Germany and was its national chair (*Bundesvorsitzender*) between 1952 and 1963. See Böhm, *Hitlers Vasallen*, 23. For a more detailed investigation, see Böhm's biography of Zillich in ibid., 60–77.

78 Fabritius, "Selbsthilfe."

79 A particularly enthusiastic review of the meeting and its prospects is offered by Scheeser, "7 Juni und 'Selbsthilfe'!"

80 Schunn, *Weg und Feinde der NEDR*, 38.

deserve to be destroyed in the coming struggle for survival."[81] Calling for membership numbers to be doubled,[82] 1931 was to witness the singularly most important shift in the movement's approach to regeneration, when its mood turned from lobby work to an outright struggle for political power.[83] The failure of the Self-Help's 1929 propaganda campaign to "clean out public life" (while still professing to abstain from daily politics),[84] the ongoing radicalization of the movement's perception of the urgency for national salvation (having come a long way from the early demands for urban vegetable patches and orchards), and the increasing dissatisfaction with the lack of tangible outcomes created a climate in which many of the Self-Help's members, and especially the newer and younger factions, demanded action.

The movement's rapid numeric and geographic expansion, paired with the ideological diversification and radicalization inherent to such a significant shift in its makeup, increasingly necessitated the formulation of a clear agenda—a task entrusted to Wilhelm Schunn, who subsequently produced the Self-Help's first program or, more precisely, a twenty-five-point set of demands addressed to the Saxon political status quo.[85] Grouped into three sections—"Basic Principles" (paragraphs one and two), "Cleaning Up Public Life" (paragraphs three to nine), and "Rebuilding" (*Aufbau*) (paragraphs ten to twenty-five)—it is a remarkably uncompromising manifesto, thoroughly suffused with not only the Self-Help's increasing focus on eugenics, but illustrating its underlying ambition to reassert the primacy of the political over religious institutions/structures, while at times most resembling a state- rather than nation-building charter.

This evolving synthesis between the increasingly overt political discourse with the various eugenic, educational, and economic policies advanced by the Self-Help since 1922 is summarized by the program's first "basic principle":

> We consider it the community's duty to ensure through appropriate organizations that every national [*Volksgenosse*] is raised in and with an idealistic and *völkisch* view of life [*Lebensanschauung*] and has enough *Lebensraum*.

81 Fabritius, "1931," 1.
82 Fabritius, "Unser nächstes Ziel."
83 Fabritius, "Kameraden," 1: "For more than half a year we have debated, considered, concerns about this or that have arisen; we thought we could manage the crisis with speeches and complaints."
84 See, for example, Jickeli, *Unser Weg zur Erneuerung*, 12.
85 See, amongst others, ibid., 15–20, where the twenty-five points are included in full.

> It, furthermore, has the duty to ensure that only those amongst us who work honestly, live simply, and think and act in noble spirit are honored—so that life in our midst may be worthy thereof. And, finally, it is given the means to work towards our nation's racial improvement [*Hebung*].[86]

After insisting that the onus similarly lay with the individual who must prove himself worthy of being German (paragraph two), Schunn proceeded to set out the Self-Help's seven core demands towards reviving the public's trust in its politicians and dignitaries, most of which threaten the most potent of measures: exclusion from the community at large. This would—reflecting Csallner's frequent demands that exclusions be used (and especially so against mixed marriages), and foreshadowing its implementation by Schunn's neighborhoods—amount to a form of internal exclusion where the shunned are banned from all national honorary positions and church ranks (most importantly the priesthood and teaching profession). It amounted to a rather long list, with the program calling for the cautioning and dismissal of all and everyone whose "family and otherwise private life is not beyond reproach. Only he who honors his and his family's honor is capable of honorably representing us" (paragraph three); of all freemasons or those with similar obligations (paragraph four); of anyone whose business practices threaten the economic livelihoods of others (paragraph six), and hence, logically, this also entailed an end to the payment of dividends by organizations with *völkisch* aims (paragraph five); of those who withhold funds needed by the community (paragraph nine); and of those teachers whose methods are not appropriate to "raise spiritually and physically whole [*einheitlich*] men, aware of being pillars of German and national culture and will" (paragraph seven). But the singularly most important demand, and one that the church, importantly, accepted in principle in treaty with the Self-Help in 1936, targeted the very priests Siegmund and Csallner had so laboriously courted, when paragraph eight called for the expulsion of "those priests from our midst that do not venerate national culture [*Volkstum*] and race as God given and worthy of all possible maintenance [*Pflege*], and we take the line that they should be held in greater contempt than one who does not honor one's father and mother."

86 This and all of the following quotations from the twenty-five-point program are quoted from Schunn, *Weg und Feinde der NEDR*, 20–23.

The final fifteen points all revolved around the form and function of internal regeneration agendas, with the first asserting essentially Csallner's definition and advancement of the "valuable" large family that the nation so urgently needed:

> We demand the family be honored and encouraged. Healthy, culturally and morally flawless parents are to be honored in and by the local community and press. Our national organizations' [*Körperschaften*] electoral practices are to be changed in such a way as to give the voter representing only himself one vote, while those representing their family are given so many votes as the family has underage members—so that for any decision each, even the very youngest of our nation's members, will have a vote. In economic terms, the family is to be supported through honorary gifts for high birthrates, through preferential treatment and economic advancement in every way possible, insomuch as practical limits and the avoidance of damage to others allow.

Saturated with eugenic ideology, and particularly so with Csallner's specific brand thereof, this is a remarkable set of demands, given that it not only involved profound changes to the Saxon political system, but placed a qualitative pronatalist agenda at the very core of its ideological and economic discourses. However, given the context discussed here, the two perhaps most interesting facets are, first, its reiteration of a voting scheme Csallner had floated in 1929,[87] and second, that it inadvertently sided with Csallner's approach to the dysgenic crisis rather than Siegmund's—in other words, the former's insistence on the primacy of increasing the nation's qualitative reproductive rate as a means towards regeneration and territorial consolidation surpassed the latter's reversal of the equation which argued that expanding Saxon *Lebensraum* was the precondition to raising fertility levels.

The *Lebensraum* motif was nonetheless omnipresent, and paragraph eleven demanded the nation's economy be regeared towards its expansion, a program to be partially funded through private savings schemes—a practice now defined as a moral imperative. Otherwise demanding the encouragement of local markets for local produce (paragraph twelve), the need for continuous training courses for various professions (paragraph thirteen), and social welfare agendas for the urban skilled labor market (paragraph

87 Csallner, "Geburteneinsturz und Familienstimmrecht," 1–5.

fourteen) and amongst the peasantry (paragraph fifteen), the Self-Help's program also demanded a significant expansion of medical provisions generally, and improved accessibility for the poor (paragraph sixteen), insisted that while periodicals may be private enterprises they were nonetheless understood as public property (paragraph eighteen), and expressed its "hope" that the Saxon school system may encourage the popular understanding of the nation as "one large family" (paragraph nineteen). There is only one paragraph to explicitly and solely deal with the economic ideology the Self-Help had been founded upon, namely paragraph seventeen, which unequivocally demanded that "our financial institutions eradicate even the last vestiges of capitalism." And, finally, reflecting its ambiguous relationship with the church(es), paragraph twenty clearly asserted the supremacy of political bodies (after demanding everyone actively further Christendom), by delegating the church to the role of an assistant, whereby it would not be "an end unto itself but a honorable institution preserving and furthering religious life as an important part of our national culture and as a service to everyone." This, again, is a rather remarkable stance towards the Protestant church[88] given its omnipresence in the Saxon everyday life, which it had effectively guided (or even governed) for decades in anything but name.

Having so clearly asserted the primacy of political over religious bodies, the program continued to demand the creation and empowerment of essential institutions to cement its authority: an employment office to control the supply and demand of labor in certain professions (paragraph twenty-one), national arbitration courts (*völkische Schiedsgerichte*) to resolve internal differences rather than having to resort to Romanian state courts (which would both cost money and could, under certain circumstances, damage Saxon interests) (paragraph twenty-two), as well as national disciplinary courts (*völkische Disziplinargerichte*). This particular set of structural demands, in any instance, can certainly be interpreted as tantamount to setting up a Saxon specific judicial branch to complement a redefined political body increasingly resembling a legislative (with the future NBs as its executive). Even if one were not to consider this a state-building (rather than a nation-building) agenda, the program's final demand (paragraph twenty-

88 Which, in fact, was held partially responsible for the leadership crisis due to its perceived inability to adapt to the interwar realities. See "Die sächsische Pfarrerfrage."

four) essentially called for a centrally guided battle against perceived state-official injustices, by having the cost of court cases of "public interest" charged to the public's purse. And so the program concludes (paragraph twenty-five) by calling upon every Saxon to be noble and heroic in spirit and live by the Self-Help's three-line mantra: "Public good before self-interest! All for one and one for all! To each his own!"

The Self-Help had benefited from the profound societal dissatisfaction with the form and function of Saxon politics, increasingly perceived as alien to the real trials and tribulations of "everyday life," a growing sense of failure that left Saxon politics effectively paralyzed by the mid-1920s.[89] It is hence little surprising to find the Self-Help assuming ever more aggressive and defamatory positions against the political establishment, its inability to respond to the growing impact of the Romanian nation-building project on Saxon economic and national interests "externally" (most notably the church and its schools), and its failure to formulate and pursue regenerative agendas "internally." While Saxon periodicals are awash with rants against the status quo's failures, a particularly interesting example of the Self-Help's perception of a historic injustice aggravated by the even more unjust realities of life in Romania is offered by Wilhelm Schunn, when he rages: "Now, when our nation was ten times more in need of protection and guidance than before the war, our leadership failed completely. Nobody seemed to care and deal [*kümmern*] with the self-inflicted daily loss of national strength and *Lebensraum*. One just thought of one's own benefits and losses. [. . .] A general lack of will, yes—personal depravity set in."[90] Raving against the politicians seen to have had "sold out" their nation through "shady deals" (*Kuhhandel*) with Romanian political parties and "voting pacts" (*Wahlpakte*), and insisting Saxon politicians had anyway let the state walk all over them, Schunn argued that "in this hopeless and completely dead-end situation, which no one but us is trying to change with serious means, in which parties everywhere have no real will, and want nothing other than to defend their own power, burst our National Socialist Self-Help Movement of the Germans in Romania to the horror of all

89 A good example of how the Self-Help labored the argument that Saxon politics had lost the public's support is offered by Rr, "Zur Krise in unserem Volk: Teil I & II."

90 Schunn, *Weg und Feinde der NEDR*, 4–5.

those needy-of-rest [*Ruhebedüftigen*]."[91] Burst a movement that had clearly written generational change on its banners, proclaiming "they call us disturbers of the peace [*Friedensstörer*], but we don't care because we belong to a new age, are of the fighting breed [*Kämpfertypus*], and have no time for bourgeois [*bürgerlich*] afternoon naps."[92]

This willingness to actively engage with political structures significantly increased when Waldemar Gust and his group of supporters in Kronstadt joined the movement in spring 1932. After the 1935 rift, Jickeli described their impact as follows:

> Simultaneous with the new organization and men that joined, a rift came into the movement. Next to the national-political [*volkspolitischen*] workers of previous days now stood the Kronstädter politicians [*Nurpolitiker*], who in heavy battle with conservative powers had learned to see the world with hate and distrust. But they also harbored this distrust against their new comrades, and that, from the very beginning, destroyed the heartfelt comradery that until then had existed amongst the circle of friends around Fabritius.[93]

IV. The NSDR Victorious, 1932–33

The cumulative impact of this increased membership—in part due to the movement's geographic expansion beyond Transylvania, in part due to the integration en masse of preexisting groups—had a substantial radicalizing effect on the Self-Help's ideology and approach to politics. The process culminated at the movement's ten-year anniversary rally on May 21 and 22, 1932,[94] when the Self-Help arrived at the fateful decision to actively pursue political power by founding a political party, the National Socialist Self-Help Movement of the Germans in Romania (NSDR).[95] The NSDR's pro-

91 Ibid., 5–6.

92 Ibid., 6.

93 Jickeli, *Unser Weg zur Erneuerung*, 22. Before the rift, Schunn had been rather more enthusiastic about Gust's influence when he wrote that: "The [movement's] gradual infiltration of the difficult Burzenland is largely his work. And he, apart from the Burzenland, has convinced the movement of the need to fight politically—which he organized and led from the front. His absence from this battle is unimaginable." See Schunn, *Weg und Feinde der NEDR*, 48.

94 See the program of the anniversary rally in Fabritius, "10 Jahre Selbsthilfearbeit."

95 Representing the movement's expansion into the other German settlement areas in Romania, the NSDR was divided into five regions: Colonel a.D. Karl von Möller became the Banat *Gauleiter*, the engineer Edwin von Landwehr that for the Bukovina, and Dr. Johann Wagner for Bessarabia, while the Old Kingdom only had disjointed local groups, and hence no regional superstructure. See Schunn, *Weg und Feinde der NEDR*, 32.

gram (published together with the draft national program in August 1932) adopted the 1931 twenty-five-point program as its core platform, but also introduced the *Führerprinzip* (interestingly enough, though, the party's leader was to be elected directly every three years, but his power was absolute while in office, including the ability to change the program).[96]

The Self-Help's division into an economic and a political wing[97] that declared war on the status quo not only widened the divide between those "old fighters" who had tirelessly insisted it was a national political "movement" rather than a "party," but symbolized its ideological adoption of the German National Socialist franchise beyond question, and its organ, the *Selbsthilfe*, soon changed its name to *Ostdeutscher Beobachter*. Still, despite this landmark shift, many Self-Help—now NSDR—members refused to accept their reinvention as a party, and so we find the likes of Wilhelm Schunn pondering:

> Have we now become a party? Yes and no! We demand more discipline than any other previous party and act as a party in public. So let them call us one. But we do not want to reach party-political goals on the backs of others, we want to be the altruistic forerunners of the future, so much yearned-for national community of the eight-hundred-thousand built on mutual help. Our public appearance is defined by our ideology. We are the organized wing of an unstoppable popular movement that will completely change the German nation. And that is why we are in our nature anything but a party.[98]

Despite Schunn's apparent dislike of party politics the NSDR was a political party with political goals, the first and foremost of which was to bully the Saxon National Council into convening a Saxon National Assembly, the Sachsentag, to fundamentally rewrite the nation's 1919 national program in line with its ideological imperatives: the nation's economic, ideological, and racial rejuvenation.

The Saxon establishment's comprehensive inability to counter the Self-Help's ever-growing influence on internal discourses on virtually everything from education to eugenics, the economy, and the appropriate political

96 See "Organisationsrichtlinien der Nationalsozialistischen Selbsthilfebewegung."

97 Where the "political" domain was defined as subsuming "all of those efforts geared towards the nation's internal renewal. We include here the youth, the moral and political renewal of our nation, and the creation of a new national organization [*Volksorganisation*]." See "Aufbruch."

98 Schunn, *Weg und Feinde der NEDR*, 34.

course vis-à-vis the Romanian state is perhaps most pertinently embodied by the turmoil surrounding the fifth (and last) Sachsentag in October 1933, the sole aim of which boiled down to agreeing upon a new national program to replace that of 1919—one which the Self-Help in particular deemed not merely a failure entirely unsuitable to contemporary challenges, but fundamentally flawed due to its focus on Saxon demands of the Romanian state (rather entertainingly referred to as "foreign policy") at the expense of Saxon internal regenerative agendas.[99] Following the joint demand released by the Self-Help and the Klingsor Circle in June 1931, the NSDR was the first to publish a draft program in the autumn of 1932, authored by Schunn with Karl Hermann Theil, and based upon the Self-Help's twenty-five-point program. This was followed by Klingsor's program shortly thereafter, while the National Council consistently dragged its feet.[100] Eventually, it did produce a draft program, but it offered "not much more than a literal adoption, partially watered down, rearranged, and dreadfully distorted interpretation of the Self-Help's draft program."[101]

While the Sachsentag had eventually been scheduled to convene in late 1932, the date was moved to May 1933, before being postponed indefinitely two weeks prior to the event by the National Council's leader, Dr. Schnell, who insisted he had received credible (if unnamed) evidence that the NSDR had planned to hijack the event through "undignified" behavior—an untenable risk to the fragile relationship between the Romanian and Saxon populations.[102] With the event thus postponed, the National Council then set up an "arbitration commission" tasked with investigating the NSDR's plans and with negotiating any remaining issues.

Clearly dismayed by what it deemed an unacceptable diversion, the NSDR convened its own National Day (*Volkstag*) in Mediasch on November 6, 1932, as a public display of opposition to the National Council, which it repeatedly condemned as a ruinously degraded "election machine" bereft

99 Ibid., 39.

100 Ibid., 38–39.

101 Ibid., 39.

102 Ibid., 41. Two particularly vocal responses to Schnell's decision are offered by Gust, "Zur Verschiebung des Sachsentages," which raved against the council's lack of preparations (which Gust interpreted as unwillingness); and K., "Kampf," which was even more bloodthirsty, exclaiming: "The National Council has declared war against our Self-Help movement in its last meeting, and war there will be." Both have a similar subcurrent, namely that the council had deliberately and unjustifiably cancelled the Saxon Day in order to defend its political power, while the latter already predicted the arbitration committee's failure.

of practical internal policies. Fabritius, for his part, read a declaration reiterating the movement's affinity with German National Socialism: "Stemming from the realization that race and national culture are God-given goods—they are to us a holy gift and duty—and realizing that selfishness and greed are veiled foreign ideas leading to Bolshevism, this national convention sees it as its noble duty to declare itself to the spirit of Adolf Hitler's teachings."[103]

The Self-Help's decision to switch its approach from lobby work to political struggle also resulted in the status quo's growing opposition thereto. While, again, the level of opposition is considerably lower than one would expect given the NSDR's radical ideology and clear affiliation with German National Socialism, it was anything but tame. Apart from sporadic local protests by individual priests and educators, the omnipresent "malevolent Jews" and the press they supposedly controlled, as well as the much maligned Marxist Gustav Ziekli in Bistritz, there was little organized resistence to the Self-Help. That is, apart from two groups that rose to challenge a Self-Help so clearly on the ascendancy in a flurry of publications and heated debates. This alliance, known as the Iron Front (*Eiserne Front*), consisted largely of Rudolf Brandsch's Hermannstädter Bürgerabend[104] and Dr. Deppner's political following in Kronstadt (and their organ, the *Kronstädter Zeitung*).

The fifth, and as Fabritius had incessantly demanded, last Saxon Day to convene in the wake of such bitter political turmoil, was held on October 1, 1933, and was beyond doubt one of the most fateful watersheds in Saxon pre-1945 history—not to mention its historiography. There can be no doubt that it marked an equally surprising and sweeping triumph for the NSDR that had suddenly found itself without serious opposition and empowered not merely through the convention's adoption of its diluted national program, but the fact that it was re-radicalized through the successful addition

103 Schunn, *Weg und Feinde der NEDR*, 40.

104 Jickeli, very interestingly, argued that Brandsch had initially been sympathetic to the Self-Help's agenda but that their much proclaimed aim of battling the freemasons (of which he was one) had caused the collapse of earlier attempts to integrate him and his group as one of the movement's "local branches," while the bitter feuding that followed had been caused by the attacks of the movement's own younger factions. See Jickeli, *Unser Weg zur Erneuerung*, 23. Schunn, for his part, insisted the Bürgerabend was in fact not a political party but the "professional association" (*Standesvertretung*) of merchants and craftsmen in Hermannstadt and that, as such, it had no role in Saxon politics. See Schunn, *Weg und Feinde der NEDR*, 44.

of previously revised or removed articles and their terminology.[105] In conjunction with the November "national elections" that catapulted them into the Saxon National Council, these two months were to mark the political empowerment of the eugenic discourse, as well. How did the NSDR's 1932 draft compare to the 1933 legislated version, and what amendments was the movement able to introduce during the Saxon Day towards "strengthening" its language and aims?[106]

The degree to which the NSDR's draft constitution wanted to mark a revolutionary break with the fourteen years of perceived political decline is perhaps nowhere stated as clearly as in its introduction: "The old one [national program] of November 6, 1919, actually does not deserve this honorable title, as it includes nothing but demands towards the state. On the other hand, it characteristically knows of no tasks directed inwards, tasks to be achieved for its nation, exactly like our leadership, which since then knew nothing of such tasks either." Unsurprisingly, the official *Volksprogramm* sported a rather more conciliatory preamble, calling upon God (who appears only once in the NSDR draft, in reference to race as a divine gift) to aid in the Saxon quest to "create an ever-more-perfect community" and "to fight for justice and our distinctiveness [*Eigenart*], for us and our descendants."

The first sections of both programs subsume five "basic principles" (*Grundlagen*), the first of which underlines how racial hygiene and a biologically defined sense of self had suffused the Self-Help's ideological discourse when its draft reads: "We declare ourselves part of the racial and cultural unity of all the world's Germans, who constitute a single nation." The official version scrapped the reference to race, while preserving the pan-German view of a globally united German nation. And while the following four paragraphs are in many ways largely the same, a few crucial issues do emerge. So, for example, the official program's third paragraph proclaimed that "the totality of the Germans living in Romania constitutes a single national and

105 In Schunn's words: "The Saxon Day clearly and unequivocally proved that our opponents want nothing. The official leadership had already adopted our 'national program,' if terribly watered down. By and large, all we tried to do at the Saxon Day was to make the wording tougher. Our opponents just sat there, dispirited." Ibid., 51.

106 All of the subsequent quotes referring to these are taken from "Das sächsische Volksprogramm" and *Volksprogramm der Siebenbürger Sachsen*. For further translated extracts from the 1933 national program, see Georgescu "Transylvanian Saxon Eugenics."

political unit, with duties and rights internally and externally." A subtle, if all the more intriguing change to the National Council's proposed program had been introduced, and accepted by the Sachsentag, by Hans Wagner, who had wanted the "duties" placed before the "rights," as he argued that National Socialism by nature demanded duties before conferring rights.[107] A final divergence of interests appears between NSDR's fifth paragraph and its official manifestation in paragraph four. The draft had demanded that "the national community of the Germans in Romania has above all else to ensure that every national [*Volksgenosse*] is raised to an idealistic and nationally oriented view of life [*Lebensanschauung*] and has enough *Lebensraum*," before concluding that it, further, "has to work with all its energy towards our nation's racial elevation [*Hebung*]." Again not surprisingly, the official national program had been watered down when it stated: "The national community has to ensure that every Saxon is raised to a national [*völkisch*] and Christian view of life. It has to create *Lebensraum* and sufficient job opportunities for its members, to strive towards social equality and—through education, enlightenment, and influence—to awaken and maintain the conviction that the German nation is a God-given unity, and every comrade is our blood-brother for whom we are all responsible."

Turning its attention to "demands of the state," both the NSDR and the officially legislated versions of the Saxon national program are, in essence, the same when they demand that the Romanian state implement the *Karlsburger Beschlüsse*; that it should further cultural autonomy (including language) and political participation; allow the minority to tax itself and maintain an independent school system (funded in part by the state), along with a host of other rights and freedoms. The list of demands becomes truly interesting, though, when it discusses economic policy, and even ventures to request reparations for "injustices suffered" at the hands of the Romanian state, with the official version insisting that: "We demand state compensation for the social and economic damages inflicted upon us since 1918, such as the agricultural reform, agricultural loan conversion, etc., and simultaneously, effective protection from any form of economic disadvantage." Utterly astonishing as this demand may be, the Saxon sense of victimization is further highlighted by paragraph twenty-four, venturing to demand that

107 "Der Sachsentag im Zeichen des Sieges der NSDR."

the state battle corruption through the depoliticization (a rather interesting choice of words) of its civil bureaucracy and judiciary. On a rather peculiar note, the NSDR's draft also demanded (paragraph eighteen) that the Romanian state's foreign policy should be reoriented towards Germany—and while wholly unrealistic, this nonetheless reflects the movement's belief that a National Socialist Romania would not only benefit them, but epitomize their role as mediators between the "motherland" and the "fatherland."[108]

That said, the program's most contentious section was its paragraph fourteen, the "Guidelines for Shaping National Life" ("Richtlinien für die völkische Lebensgestaltung"). The NSDR's draft had copied these from its 1931 twenty-five-point program, reiterating its calls for the exclusion of, amongst others, those with anything other than flawless personal lives, those with "damaging" international connections (notably freemasons and Bolsheviks), those whose business practices threatened the livelihoods of others, those teachers deemed "unsuitable" and, of course, those priests who did not value race and (its) culture as God-given gifts. Most of these stipulations did find their way into the official national program, if as recommendations rather than accompanied by threats of exclusion. But how many of the movement's eugenically framed demands was the NSDR able to insert into the official version, and in what wording? Quite a lot, considering we find the first paragraph demanding a popular willingness to sacrifice fiscal and personal gains for the nation, and paragraph two stating: "The maintenance and expansion of *Lebensraum* is one of the national community's fundamental duties. Our nation's peasant foundation must, as its natural source of strength and renewal, be preserved, solidified, and extended." But the singularly most intriguing paragraph is the third, declaring "the questions of national health, hygiene, racial hygiene, and population growth constitute a considerable portion of our national organizations' work agenda. Apart from public enlightenment, this work will be conducted through institutions of national health, especially through the creation of national hospitals [*Volksspitäler*], the preferential treatment of large families, and by battling child limitation [*Kinderbeschränkung*]" (although these are clearly watered-down versions of the twenty-five-point demands articulated in paragraphs

108 As this argument is frequently encountered, also see Weber, "Ein offenes Wort," who claims National Socialism would "have a fruitful [*befruchtend*] influence on the Romanian nation, and thereby continue and fulfil the German nation's historic mission to be a mediator in Romania in the best possible way."

ten and eleven quoted above). That said, paragraph four logically continued to postulate the family's national importance and offered a rare perspective on gender relations as defined by this eugenic context when it professed "the highest honor and calling for women is motherhood [*Mutterglück*]."

In terms of its economic policies, the NSDR similarly succeeded in getting two of its key demands adopted by the Sachsentag. First, the convention accepted Jickeli's proposal to reintroduce one of the movement's oldest desires—namely, "the extinction of even the last vestige of capitalism"[109] from Saxon fiscal institutions—into paragraph eight (from which it had been stricken). Second, and following Dr. Wagner's proposal, the passage demanding that the economy further the path "from local producer to local consumer" was reintroduced into paragraph nine.[110] With regards to the church and its place in Saxon society, the *Volksprogramm* also featured an article enshrining the supremacy of politics, declaring "the national community reserves the right for itself to influence the effectiveness of church and school in all national matters" (paragraph seven).

But the two most significant battlegrounds with conservative convention delegates revolved around two of the Self-Help's flagship policies: enshrining "voluntary youth work" as mandatory for anyone pursuing a church or national honorary office, and the democratization of Saxon electoral practices. The NSDR draft program, under paragraph thirty-eight, had demanded that "mandatory work [*Arbeitsdienstpflicht*] is to be introduced as the national education's crowning achievement." Despite significant opposition to this paragraph, the NSDR's Gust and Staedel fought for it bitterly,[111] and following an amendment to their amendment by Roth, the measure went through by a margin of 299 to 216 votes, thus rendering voluntary work mandatory for any Saxon born after 1915 (paragraph six).[112]

Pushing through its second set of key policies, the equally radical reforms to Saxon political structures, was only moderately successful. Paragraph forty-four of the NSDR's draft program had called for the Saxon National Council[113] and the Saxon Day's delegates to be elected directly by popu-

109 "Der Sachsentag im Zeichen des Sieges der NSDR," 2.

110 Ibid.

111 See, in particular, WR., "Was ist Arbeitsdienst?"; and "Was wollen wir auf dem Sachsentag."

112 "Der Sachsentag im Zeichen des Sieges der NSDR," 2.

113 They wanted to change its name to Regional Council, in line with the movement's overall pan-German ideology, and the title of its leader from Chairman to the historic Komes.

lar vote (*Urwahl*), with proportional representation. The NSDR won its bid with regards to the latter, on Gust's initiative, but his second motion, supported by Jickeli proposing the National Council's reform, was narrowly defeated by 282 to 240 votes. A third reform to Saxon voting practices put to the Sachsentag demanded proportional representation in both rural as well as urban areas (the National Council had spoken in favor of the latter only)—a demand Waldemar Gust astonishingly succeeded in introducing into paragraph eight of the official program's section on Saxon "national organization." And while this reform ultimately democratized historic Saxon electoral practices, the pursuit of democratic electoral means was a necessary measure if the Self-Help wanted to be voted into political office. As a commentator noted in the *Ostdeutsche Beobachter*, the movement had been forced into the guise of a political party in order to pursue the political power they could only ever achieve through a proportional vote—a problem illustrated by the Hermannstadt district council (*Kreisausschuss*), in which the "Self-Help with its 700 members had the same number of delegates as the city's seven freemasons."[114] Put more combatively:

> New ideas can only be represented by new men. Because the current system prevents the ascension of new, unused energies [*Kräfte*], we have to push through a new electoral system at all costs. We are hence never going to budge from our demand to introduce a proportional vote for the National Council in both cities and rural areas alike. [. . .] Yes, does anyone really think that at an *Urwahl* only so much as half of its current members would be reelected? No, we are certain that based on a proportional vote the new National Council would have an entirely different composition than it does today.[115]

The much anticipated Sachsentag had amounted to nothing less than an astonishing victory for the NSDR. The movement thoroughly revolutionized the Saxon national program in line with its core ideological values, and subsequently legislated eugenic and race-hygienic policies, even if frequently in a compromised form. The comprehensiveness of the movement's success, though, is not only evident in the official adoption of their draft national program, but in the NSDR's surprising ability to amend and

114 "Vor der Entscheidung," 12

115 "Durch den Sachsentag," 1–2.

reedit certain passages therein. Jickeli captured that movement's exuberant mood perfectly when he wrote: "The Saxon Day brought us an overwhelming victory. Although the bodies that sent the delegates consisted largely of our enemies, we managed to take the Saxon Day's leadership into our hands and corner our enemies into the role of a checked [*schächlich*] opposition."[116] However, revising the program was only the first step, a launchpad for the Self-Help's bid for the political supremacy that it achieved a month later.

116 Jickeli, "Kameraden der NSDR," 3. See also "Ausblick nach dem Sachsentag."

CHAPTER V

SAXON FASCISM IN POWER, 1933–40

The NSDR's astonishing ability to impose its national program had a profoundly symbolic function in marking, as the Self-Help's organs had so frequently quoted Fabritius insisting, the beginning of a new age, of an entirely new understanding of national identity.[1] But it was also only the first of three political victories that the NSDR enjoyed in rapid succession. A month later, on November 4/5, 1933, the elections to the Saxon national bodies handed the movement the Saxon National Council, and Jickeli became its new president (in Hermannstadt they defeated their historic rival Rudolf Brandsch and his Bürgerabend,[2] having gained sixty-two and seventeen percent of the vote, respectively).[3] Thus swept into the very political office they had battled a month previously, Schunn delighted: "With that, the National Socialist Self-Help Movement of the Germans in Romania assumes the responsibility for Transylvania's future development. This development is in particular going to be the National Socialist remodeling [*Umgestalung*] of the nation's entire life. Our energies are in the first instance those of internal regeneration [*Aufbau*]. May the victory bring blessing not only to the old settlement areas of the Saxons, but to all of the other younger settlements as well."[4]

---

1 Amongst the plethora of pre-October journal articles advertising the Sachsentag, see Gust, "Letzter Appell"; and Fabritius, "Deutsche Volksgenossen."

2 Rather entertainingly (and illustrative, considering the generational shift underway then and in the early 1930s), the following article reprinted a scathing criticism of a 1914 Bürgerabend meeting published in the *SDT* to demonstrate that in its heyday it had displayed worse behavior than the NSDR: "This assembly's proceedings cannot possibly be called dignified. When slogans such as 'shut up' or 'throw him out' are heard at a Saxon electoral meeting, then this is a sad indication of just what low point our political life has sunken to." See "Bürger-Abendrot," 4.

3 Jickeli, *Unser Weg zur Erneuerung*, 25; and "Ein Ausschnitt aus den bisherigen Wahlerfolgen."

4 Schunn, *Weg und Feinde der NEDR*, 53.

The Self-Help was inherently a pan-Germanic movement, one that not only aspired to salvage and regenerate the Saxon nation, but that eagerly advertised the opportunities to unite all of the various German settlement areas in Greater Romania. Fabritius repeatedly declared that the 1933 Saxon Day would be the last, and that the next constitutional assembly would be a German Day, when all of the country's Germans would agree to a communal national program. Not only does this signify a historically remarkable conclusion to the same *klein-sächsisch* vs. *all-Deutsch* debates to have raged throughout each of the three previous generations' political discourses,[5] but it also necessitated the Self-Help's active support of Romanian attempts to ward off irredentist claims. This point is perfectly illustrated by Jickeli's inaugural address as Transylvania's Regional Leader (*Gauleiter*) in Hermannstadt, on June 7, 1933: "Romania's borders are also holy to us, in that Greater Romania has unified the majority of Southeastern Europe's German settlement areas. Our national comrades all live in endangered border regions. [...] If we demand a revision of Romania's borders, then it can only be to demand that parts of the Banat currently ruled by Yugoslavia are united with Romania, so that the German comrades living there may be added to our national community."[6]

1. The Self-Help's Various Forms and Formats, 1933–34

On its path to political dominance of all German enclaves in Romania, the Self-Help underwent two more fundamental upheavals: the first came in the

5 See, for example, Möckel, "Kleinsächsisch oder Alldeutsch?" This piece offers a rather interesting summary of the essentially four historic intergenerational public debates on Saxon identity and its shifting location within the wider pan-German discourse: first, the "old" vs. "young Saxons" dispute following the 1867 *Ausgleich*, where the younger generation believed cooperating with Hungarian rule might be more beneficial than embarking on an overt struggle to defend historic rights; second, the conflict between the "Black" and "Green" movements that surrounded the 1896 Saxon Day; third, the *klein-sächsisch* vs. *all-Deutsch* debate that raged unto the First World War; and, on a final note, Möckel perceived the interwar conflict between the "movement for renewal" and its conservative adversaries as a fourth intergenerational struggle for political power.

6 Quoted from Schunn, *Weg und Feinde*, 78. The speech was also published as Jickeli, "NSDR oder 'Eiserne Front.'" It also touched upon another important theme, namely the movement's relationship with the church when Jickeli, his first act as *Gauleiter* having been to visit Bishop Glondys, insists that Saxon national regeneration was "anchored in Christian ethics," and that the NSDR would place itself at the "church's disposal" when it came to ensuring that church taxes were collected (by threatening those who refused with exclusion). Again, though, and despite the rhetoric, the underlying message that the political was carving a space for itself as guardian in the religious—or, more appropriately, structural—domain is apparent.

form of a government ban dissolving the NSDR on November 29, 1933. This led, ten days later, to the formation of the National Movement of Renewal of the Germans in Romania (NEDR)[7] which, second, was banned as well, on July 4, 1934.[8] Nonetheless, it is remarkable that neither ban appears to have hampered the movement's ability to maintain its coherence or exert its political influence. In effect, they had been issued too late given that the Self-Help come NSDR come NEDR controlled both the Saxon and Bessarabian National Councils[9]—although it certainly did impact the movement's progress in the Banat and Bukovina.[10]

The NEDR's dissolution certainly tore upon a fundamental rift between the movement and Bishop Glondys, whom it held directly responsible for the Interior Ministry's ban. The first bout of outright struggle between the Self-Help and the then *Stadtpfarrer* of Kronstadt, Viktor Glondys, was in autumn 1931, after his delivery of what became known as the "Good Samaritan Sermon" on September 6, 1931. Given the Self-Help's deeply eugenic approach to national regeneration, Glondys must have been aware that his frontal assault on the "race cult," and insistence that racial hygiene was diametrically opposed to the Christian Samaritan nature (concluding that "Christianity will overcome racial hatred through the commandment of brotherly love, which transcends the barriers of one's own race"),[11] would evoke the Self-Help's vocal opposition. With the dispute eventually declared "resolved" in early 1932,[12] a confrontation with the church proper followed shortly thereafter, and centered on the election for a new Saxon bishop to succeed the retired Friedrich Teutsch. In short, the Self-Help had petitioned the church in August 1932 to postpone the election until after that of a new Landeskonsistorium in early 1933, arguing that to have a leaving "parliament" (from which the Self-

7 Reflecting the increasing significance of the eugenic/race-hygienic discourse, NEDR membership applicants had to declare: "I hereby declare becoming a member of the National Movement for Renewal of the Germans in Romania. I am of German-Aryan descent, am not a member of a freemasonic lodge, or of any other secret society, and do not belong to any Marxist (social democrat or communist) organization." See NA Sibiu, CGR, A.I.1. Album A, p. 63.

8 Fabritius, "An meine Kampf- und Gesinnungsgenossen!"

9 See "Ein Ehrentag des Deutschtums in Bessarabien," 4, which exclaimed: "On the old system's ruins the NEDR now stands victorious and ready to take over responsibility for shaping the German destiny in this settlement area." See also Gust, "Das bessarabische Deutschtum unter neuer Führung"; and Schunn, *Weg und Feinde*, 95–96.

10 Jickeli, *Unser Weg zur Erneuerung*, 41.

11 Glondys, "Samaritergeist."

12 See "Zum Abschluss."

Help was as good as absent) elect a new bishop as its final act violated both the spirit of the time and the church constitution.[13] But the church pressed ahead with electing its new bishop, and the Self-Help responded by setting up its own candidate, Father Wilhelm Staedel, who lost to Glondys.[14] Subsequent to the NSDR's rise to power at the Sachsentag, Glondys published a scathing attack on the movement in the form of a flier entitled "To Clarify the Situation, a Word to All Saxons" ("Zur Klarstellung der Lage, ein Wort an alle Sachsen"), which quoted extensively from the movement's "members' manual" (*Dienstbuch*). The resultant flurry of accusations in both directions eventually culminated in the signing of an "agreement" between the NEDR and church on March 16, 1934, according to which the NEDR was obliged to suspend enforcing the *Führerprinzip*, one of its central political platforms, when it came to members employed by the church.[15]

The NEDR clearly chose to profile this agreement as evidence of their willingness to negotiate, and Jickeli for his part proclaimed "bells should have rung throughout the Saxon lands—future days will speak of this historic event,"[16] in a speech given to the NEDR national assembly in Mediasch on March 18, 1934. Nonetheless, the agreement marked a clear defeat, as the movement accepted the church's ultimate authority over its employees:

> The NEDR's leadership abstains from applying the *Führerprinzip* in terms of its ability to issue binding orders to members of the Protestant National Church AB in Romania, in all matters covered and regulated by this Church. In those instances where the NEDR leaders are themselves members of the Protestant National Church AB in Romania, they accept the church constitution [*Kirchenordnung*] and regulations as binding. All of the employees of the Protestant National Church AB in Romania are to deliver written declarations stating that they do not recognize, and consider invalid, those orders contravening the agreement with the NEDR presented here.[17]

13 Fabritius, "Wer soll den neuen Bischof wählen?"

14 See "Aufbruch."

15 The Self-Help repeatedly insisted, though, that this agreement had merely formalized the movement's repeated assurance it would not undermine the church, that "this agreement's quintessential core lies with the stipulation that the NEDR should commit itself to not issue mandatory orders that would force its members to violate key tenets of the faith or of the legal basis for the church and their oath of office." "Frieden mit der Kirche," 3. See also "Vereinbarung zwischen ev. Landeskirche AB in Rumänien und NEDR."

16 "Die grosse Volksversammlung der NEDR," 5–6.

17 NA Sibiu, CGR, A.I.1. Album A, pp. 71–72.

Despite this rapprochement, Glondys's 1932 flier was considered to have offered the Romanian state the excuse and evidence it needed to disband the NEDR, especially as the ministerial decree justified its decision with the movement's "hefty" public appearance and polarization of Saxon society.[18] This was certainly the view of commentaries published in the German *Völkischer Beobachter,* which led an infuriated Glondys to formally protest their assertions at the German embassy in Bucharest. But the flames were stoked ever higher when the "former leaders of the NEDR" Fritz Fabritius, Dr. Alfred Bonfert, Gustav Prall, Herwart Scheiner, Dr. Waldemar Gust, Dr. Albert Dörr, Karl Hermann Theil, and Dr. Michael Prall put their names (interestingly, Csallner, at this point still a priest, did not) to a flier cynically entitled "To Clarify the Situation: An Answer to All German Comrades."[19] It was an unequivocal declaration of war, whose sole purpose was to insist that "the signatories endorse the position of the German press," while offering several examples of how Glondys had attempted to undermine the National Socialist cause, including the aforementioned "Good Samaritan Sermon." These included Glondys's attempts to have Jickeli declare the NEDR disbanded at the National Council's first meeting in January 1934; that he had prevented church employees from joining the movement; that his flier had distorted the NEDR's work and influenced the Interior Minister; and that Glondys had anyway "rejected and fought against" the NEDR. The declaration concluded that "given the stance Bishop Glondys has assumed towards National Socialism then and to this day, we consider his work within the German national community as incompatible with the honour of Germanness [*Deutschtum*]." As history would have it, Glondys was indeed pushed out of office by Schmidt, and replaced with his former rival, Staedel.

On the political side of things, the NEDR had found another vocal opponent in the towering figure of Hans Otto Roth[20] and his Unity Party (Einheitspartei), which accused the movement of totalitarianism,[21] of being a political failure, and—due to its behavior—incapable of collaborating with

18 See "Die Auflösung der NEDR im Amtsblatte verlautbart."

19 "Zur Klarstellung der Lage."

20 In fact, Gust accused Roth of being the architect of the NEDR's dissolution, as his Unity Party had been formed shortly after the publication of Glondys's flier, which he suspected represented Roth's attempt to "now crawl behind the bishop's back." Gust, "Der wahre Schuldige."

21 A point to which the NEDR responded by insisting that their "totalitarian" ambitions were ideological, not political, as seen in Dr. V. C. J., "Eine Erwiderung"; and "Exekutive?!"

other parties.[22] But the "political failure" was the result of fifty-three of the National Council's delegates walking out of its first meeting with Jickeli as president, on January 21, 1934.[23] This effectively shut down the council unto their return, the dispute unresolved, on April 22, 1934 (because, as Samuel Karres's declaration put it, they wanted to "resist looming serious dangers").[24]

II. War and Peace: The National Community of Germans in Romania, 1935–40

Embattled and without an organizational superstructure, the Self-Help managed not only to hold on to power in the Saxon National Council, but to arrive at its ultimate destination on June 29, 1935, when Fabritius was elected president of the much derided Association of Germans in Romania (VDR). The Self-Help had repeatedly and with increasing intensity criticized the VDR's perceived failure to create an organic and political sense of national belonging amongst the various German settlement areas after 1919.[25] So, for example, Wolff summarily dismissed it as "nothing more than the union of German parliamentarians which discusses German issues with the government,"[26] while Hans Hockl simply called it a "still birth."[27]

This broader nation-building project to make "brothers out of brothers"[28] is perhaps most effectively summarized by Hans Hockl (who was to head the Self-Help's youth movement after the rift with Bonfert) in a two-part article entitled "The Germans in Romania Become a Nation," published in July 1933. The first part not only offers a revealing assessment of the German settlement areas' relative organizational and "national fitness," but one of the few definitions of nationhood provided by the Self-Help: "I understand a national community [*Volksgemeinschaft*] to be: a national community with a coherent political will—in contrast to a nation [*Volk*] which represents a

22 See a comprehensive criticism of Roth and his Unity Party's policies and tactics in "Die NEDR baut auf"; Neugeboren, "Rundschreiben der Gauleitung Siebenbürgen"; "Spiel mit dem Volk"; and Scheiner, "Selbst Entlarvt!"

23 "Die grosse Volksversammlung der NEDR," 5–6.

24 See "Amtlicher Bericht."

25 For example, Wolf, "Zur Konstituierung des Verbandes der Deutschen in Rumänien."

26 Wolff, *Ein Jahr Volksgemeinschaft*, 5–6.

27 Hockl, "Die Nationswerdung der Deutschen, Teil II."

28 Jekelius, "Wir wollen sein ein ewig Volk von Brüdern."

life-biological [*lebensmässig-biologischen*] term."[29] The interesting point is, of course, that Hockl operates with an entirely biological definition of "nation" rather than a historical-cultural or linguistic paradigm. The article's second instalment refines the remit of this envisioned "nation of the 800,000": "The culture [*Volkstum*] is supposed to remain, as they represent rooted national particularities [*Wesenseigentümlichkeiten*]. Apart from that, though, every tribe [*Stamm*] has to be placed under the nation [*Nation*] that has a political will, a national battlefront [*Kampffront*], and the single consciousness of being the champion [*Sachwalter*] of all Germans."[30]

The discord plaguing the Self-Help since its rapid expansion in the late 1920s and early 1930s now fully came to the fore, and split the movement into a moderate wing including Schunn, Jickeli, and Fabritius, and a radical faction led by Gust and Bonfert (which also controlled the *Ostdeutscher Beobachter*'s successor, the *Tages-Zeitung*). Electing Jickeli's successor as leader of the Saxon National Council proved to be one of the early but pivotal political encounters between the movement's two wings, one which the moderates won when Wolff beat the DVR's candidate Herwarth Scheiner with seventy to thirty percent of the vote. Jickeli believed that this defeat led to Gust's decision to stage the attempted coup during the VDR's general assembly in Bucharest, in June 1935.[31]

In a nutshell, the radicals, if only briefly, had succeeded in revising the VDR's structure by inserting a new governing council alongside Fabritius's role as president, consisting of Gust, Bonfert, and Minnich. A month later, the VDR's meeting on July 28, 1935, conclusively ended the putsch by revoking the mandates of the "three-committee" and confirming Fabritius in his office with a forty-nine out of sixty-seven vote of confidence.[32] Fabritius's hand, having survived the coup politically, was subsequently significantly strengthened with Hans Hockl's declaration of allegiance that brought the youth movement back under its control.

Ironically, the Self-Help's moderates united with their conservative opponents in order to battle Gust's new political platform, the German National Party in Romania (DVR). This new National Socialist party had initially

29 Hockl, "Die Nationswerdung der Deutschen, Teil I," 1.

30 Ibid., 2.

31 Jickeli, *Unser Weg zur Erneuerung*, 46.

32 Ibid., 49–50.

been founded in the Banat to substitute for the banned NEDR, but Gust subsequently created a second branch in the Bukovina, before in mid-July (poetically, the 14th) 1935 the Transylvanian DVR was formed and soon handed over to Alfred Bonfert's leadership. The failed putsch had a deeply significant generational connotation, particularly so due to Bonfert's central role therein. In essence, it represented the radicalized youth's attempt to embark upon its own permanent revolution, and given that he had been one of Fabritius's earliest "disciples," it was particularly damaging.[33]

III. The Mighty Pen: The 1935 National Program of Germans in Romania

With the coup thwarted, the VDR proceeded to task Wilhelm Schunn with drafting a new national program for all of the German enclaves. The revolutionary program he tabled on October 22, 1935, not only passed through this legislative chamber, but did so with a two to one majority of forty-seven to twenty-one votes.[34] So what did this new, pan-German program propose, and how did it differ from its 1933 template? First of all, it tellingly changed the organization's name from Association to National Community of the Germans in Romania (*Verband* became *Volksgemeinschaft*). It was deeply symbolic considering that the German National Council increasingly resembled a centralized federation under Fabritius's leadership, one that soon renegotiated its 1934 agreement with the church. A reflection of the new political realities, the treaty's 1936 rendition finds the church "voluntarily emphasize that it considers race and culture [*Volkstum*] as God-given values that it is, of course, duty bound to cultivate [*Pflege*]" (paragraph three).[35]

Subsuming six sections covering "basic principles," "guaranteeing func-

33 Indeed, as Jickeli—if rather polemically—put it: "Dr. Bonfert [was] like a son to Fabritius's heart. [. . .] He had frequently said, in front of Bonfert as well, that he only wanted to lead for a few more years, until Bonfert was fully matured [*ausgereift*], so he could return to his plough. That is why he, for the longest time, couldn't believe that this, his favorite comrade, would inflict the greatest disappointment of his life." Jickeli, *Unser Weg zur Erneuerung*, 46.

34 Ibid., 50; Wolff, *Ein Jahr Volksgemeinschaft*, 9. Interestingly enough, Schunn submitted a report in March 1936 stating he had written the new program during a three-month bout of paratyphus. See Schunn, "Arbeitsbericht des Beauftragten für den Volksorganischen Aufbau," NA Sibiu. CGR.D.XIII.A.227, pp. 8–10.

35 "Vereinbarung zwischen der evangelischen Landeskirche A.B. in Rumänien und dem Volksrat der Deutschen in Rumänien bezüglich der Durchführung des 'Volksprogramm der Deutschen in Rumänien,'" NA Sibiu, CGR, A.I.1. Album B, p. 1a–e. See also Wolff, *Ein Jahr Volksgemeinschaft*, 17–18.

tionality," "internal work," "external work," "national organizations," and conclusions, the 1935 national program[36] was in many ways less radical in vocalizing the movement's eugenic ideology compared to Schunn's 1933 Saxon version. It also exhibited a far greater focus on structural organization, thereby creating the framework within which Csallner's expanding LSS and Schunn's national neighborhoods were to find particularly prominent positions.

The program's "basic principles" were largely identical to those of its Saxon predecessor, if phrased more imperatively. So, for example, the first paragraph emphasized that "all of the world's Germans constitute a single nation, of which we are part," while continuing with proclamations of their loyalty to the Romanian state. That said, the program turned its gaze "inwards," summarizing its aims and ambitions thus: paragraph five unequivocally stated that "our goal and measure [*Mass*] is our nation's renewal," a purpose reiterated in paragraph seven, which defined its pan-German nation-building project—"The national leadership's internal goal is the creation of a true community encompassing all of the land's Germans, that knows no settlement area borders, in which every national comrade sees in his comrade a brother of the same blood, and which is built on comradery, mutual aid, and the maintenance of our German nature's [*Wesensart*] purity." But that this "nation of blood brothers" was anything but a loose federation is exemplified by paragraph six, which insisted that "everyone has, through himself and his financial ability, to subordinate himself under the community from whom he has received everything that makes his existence as a German worth living." The only "principle" directed "outwards" is paragraph eight, which tasked the national leadership with building "trusting" inter-national relationships (while endeavoring to secure its *Lebensraum*).

With the "nation of the 800,0000" thus defined, the second section of the 1935 program, on "guaranteeing functionality," asserted the primacy of politics over virtually every sphere of public and private life. In short, it sought to enshrine monopolies over: the individual through mandatory voluntary labor (paragraph nine) and otherwise subordination to national policies (all of which also enjoyed a six-month grace period during which pub-

36 All the quotes from the "National Program of the Germans in Romania" ("Das Volksprogramm der Deutschen in Rumänien") are from the facsimile published in Böhm, *Die Deutschen in Rumänien*, 373–89.

lic criticism was prohibited, with violations thereof defined as a "hostile attack on the community's ability to live [*Lebenskraft*]") (paragraph eleven); the press, essentially through censorship (paragraph twelve);[37] and perhaps most interesting of all, over every organization and society, apart from the church, by submitting them to the council's political will (paragraph ten). In fact, all public "officials" had to submit a written declaration (refusal to do so being tantamount to resigning), stating: "I solemnly swear to exemplarily fulfil the national program, to obey the orders given by the National Community's offices, and to use all my abilities to work towards the goal: the true national community of all these lands' Germans, built on comradery, honor, loyalty, responsibility, and mutual aid" (paragraph fifty-two).

In other words, this set of stipulations enshrined and expanded the *Führerprinzip* to subsume all German settlement areas under one centralized political superstructure, which defined its visions for (pan-)national regeneration and renewal in section three, on "internal work." Spanning sixteen individual paragraphs, the program regulated education (paragraphs thirteen and fifteen), professions and class structure (paragraphs thirteen, and twenty to twenty-three), cultural work (paragraph seventeen), legal aid (paragrapah twenty-eight), and fiscal policy, with calls on German banks to fulfil their national duties (although the customary rants against capitalism are absent here) (paragraph twenty-four). But there are also some points that merit a closer look. For example, paragraph nineteen marked the peasantry's first appearance as the fountain of national virtue in any of Schunn's programs (which is in itself peculiar given its centrality to the Self-Help's overall discourse), when it stated: "The peasantry has to be filled with the consciousness of its special value to Germanness [*Volkstum*] and its special duties. All other estates are to be convinced that our destiny is particularly dependent upon a high-quality [*hochwertig*] and rooted peasantry. [...] The peasants' vocational training is indivisible from their spiritual shaping [*Formung*], and has to be increased in all settlement areas." This "spiritual molding" is equally applicable to the youth, and so paragraph twenty-five

37 An example of what this implied is afforded by the memo of Arthur Fink, head of the National Office for the Press and Propaganda, on how to portray the outbreak of the Second World War: "We are assuming a completely subjective stance in favor of our motherland. [...] The objective informative role of our press no longer applies [*fällt weg*]." See NA Sibiu, CGR, D.XIII.A.278, p. 148. This memo was also accompanied by an article written by Fink, the first sentence of which tellingly reads: "The unavoidable has happened. The German Reich, despite the greatest of efforts, has not managed to prevent the war." Ibid, 149.

demanded they be raised as "idealistic and nationally minded German people" through mandatory work for anyone born after 1920—the participation in which was a precondition for holding any "public" offices. But that the Self-Help's new "idealistic" man needed an idealized body was warranted by paragraph twenty-seven that insisted it was a moral duty to place physical education on par with academic training, which demanded a provision of health care adequate to ensuring the nation's "racial worth."

The central role that eugenics ascribed the family is evident in paragraph eighteen, a literal reiteration of the 1933 program's use of public honors and fiscal measures to provide incentives for large, "valuable" families. There is a single, short, but all the more significant addition: "Entering mixed marriages is rejected [*abgelehnt*]." This is a peculiar word choice given that "*ablehnen*" can be translated to mean decline, disapprove, or discourage, and it is uncharacteristically open to interpretation compared to the Self-Help's customarily voracious disdain thereof. Nonetheless, the clause is particularly significant given not only Csallner's influence on the movement's eugenic mindset, but the obvious impossibility of implementing any such policy. It clearly points to a further radicalization, one that marks the onset of fascist engineering works on the eugenic fortress the Self-Help strived towards.

Turning its gaze "outwards," if only very briefly, in four paragraphs, the program reiterated the importance of "creating" (again, a rather interesting word choice given that the program was published in the wake of the NEDR's dissolution by the state) a "trusting relationship with the Romanian nation," before launching into a series of demands: the implementation of the Karlsburg Decrees (paragraph thirty-one); the inalienable right to self-taxation, language preservation, schools, etc. (paragraph thirty-two); and, finally, that the Romanian state combat (or, as the program puts it, "heal [*gesunden*]") corruption and legal injustices. Again, this rather short section exhibits one of the central truths underlying the entire Saxon fascist worldview, namely that the Romanian state is accepted as something of a malevolent force of nature damaging to, but divorced from German internal politics. Even if this is not interpreted as a state- rather than a nation-building project, Romania's apparent and almost complete absence from its programs is rather striking, if ultimately characteristic of a thoroughly self-absorbed movement obsessed with national renewal.

This multitude of agendas was to be overseen and controlled by the rather impressive apparatus outlined in the program's largest section, spanning eighteen paragraphs on "national organization." The National Community of Germans in Romania, headed by the legislative National Council of Germans in Romania (Volksrat der Deutschen in Rumänien), represented the seven regions or *Gaue*: Old Kingdom, Banat, Bessarabia, Bukovina, Dobrudja, Sathmar, and Transylvania (it is important to note that while the Banat had eighteen, and Transylvania seventeen delegates, the other regions had a combined total of nine, as the votes were distributed according to population size).[38] Reviewing its "membership criteria," the first paragraph (thirty-four) illustratively stipulates that: "Regardless of gender, every German in these lands who agrees to this program and does not exclude himself through his behavior is a member of the national community unless explicitly excluded. [. . .] Entering nationally degrading international commitments (Freemasonry, Marxism, etc.), or publicly endorsing [*vertreten*] worldviews than contradict this program's basic principles (e.g., Marxism) results in the loss of all honorary rights."

The German National Council, for its part, elected the "national leader" (as with all other posts, with a three year mandate),[39] who appointed and dismissed a number of "helpers" (*Gehilfen*), including the leader of the German Parliamentary Party and a range of "commissioners" (most resembling ministers). Amongst these we find the singularly most important post in this new national superstructure: the Commissioner for the Nation's Organic Constitution (*Beauftragter für den Volksorganischen Aufbau*), of which there was only one during the National Community's brief existence, namely Wilhelm Schunn. For his part, Schunn defined his new title literally when he insisted that "as the term 'organic' (and not organizational) shows, we wish that our national group becomes an organism, a living body [*leben-*

38 It should also be noted that the German National Council also included nonvoting members, such as parliamentarians and the "special commissioners" (*Fachbearbeiter*). Further, the number of regional votes would be halved if a region failed to pay its taxes (set according to population size and economic ability) at the end of a twelve-month grace period. Its executive committee (*Vollzugsausschuss*), though, was composed of regional representatives to the German National Council, and offered them a considerably larger vote than the council itself, with the Banat and Siebenbürgen's twelve votes to their combined six (paragraph forty-one).

39 The 1935 national program also included rather sweeping democratic reforms. So, for example, the regional councils independently decided on their size and composition, and were elected by direct vote with proportional representation (paragraph forty-four). Similarly, the city (*Ort*) and local (*Kreis*) councils were also elected, while all groups with less than ten percent of the vote where not represented (paragraph forty-five).

diger Leib] in which every cell has its tasks."[40]

The pivotal role Schunn created and assumed with this post is evidenced by the wide-reaching responsibilities it entailed, encompassing virtually everything by virtue of the incredible degree of jurisdictional overlap the organization created (and that offers such a valuable comparison to the Third Reich's equally convoluted apparatus). In short though, Schunn directed the National Neighborhood Office, which itself subsumed, amongst others, mutual aid (Neighborly Help), women's work, youth work, societies, Csallner's LSS, *Lebensraum* and the necessary funds to expand it, legal aid, the arbitration courts, national health, ideological training, and cultural policy.

Just how central the neighborhood system was supposed to become is apparent in paragraph fifty, which postulated it as the quintessential interface between politics and its body politic—one characterized not only by ethnic, but ideological exclusivity: he "who refuses to be integrated [*Eingliedern*] in a neighborhood has excluded himself from the national community." The neighborhoods were hence designed as the centripetal core of the social and economic life, one that could either embrace and support, or disgrace and exclude any given individual. In short, the neighborhood system of legislating and enforcing every aspect of German social, economic, educational, eugenic and race hygienic, medical, and ultimately political life was absolutely pivotal to the fascist bid to sieve and salvage what it deemed fit in not only the Saxon nation, but all of Romania's presumed Germans.

IV. Building a Bristling Eugenic Fortress, One Neighborhood at a Time: Wilhelm Schunn's National Neighborhoods, 1933–40

Beyond the rhetoric, the rivalries, and the relentless press feuds that fraught the years between 1935 and 1938, perhaps the most tantalizing insight into the Self-Help's attempts at translating its eugenic discourse into viable policies is afforded by Wilhelm Schunn's network of national neighborhoods. The beating heart of the Self-Help's bid to create a eugenic fortress, and the ultimate realization of practical means to Csallner's eugenic ends, they represented the executive arm of the Self-Help's increasingly centralized

40 Schunn, "Der Start zu einem Hohen Ziel," 47.

national apparatus. The degree and manner in which the movement hoped to retool these historic Saxon institutions[41] to suit their needs is illustrated perfectly by their "innovator" Wilhelm Schunn: "Given the central position that the neighborhoods have been given by our national program, and that they are tasked with the role of being the national community's executive organ [*Durchführungsorgan*] in almost all domains, it was self-evident that the neighborhoods' leader—the commissioner for the nation's organic constitution—would also be given all other non-economic [*berufsständisch*] internal [*Innenleben*] resorts to oversee."[42]

Schunn hence became something of a virtual "minister of domestic affairs," complete with his own means of enforcement. But as the Saxons did not have a state of their own, the entire system relied on these neighborhoods creating a vast bureaucratic network,[43] equipped with the only two real coercive powers they had—the *Führerprinzip* as a moral imperative at one end, and the threat of exclusion at the other.

Although Schunn's vast organization counted twenty-two individual departments by 1939, the neighborhoods have been almost entirely forgotten by historiography. This situation is aggravated by the fact that the handful of available texts have been largely rendered outdated by the recent availability of the Romanian National Archive's vast treasure trove in Sibiu, spanning almost five hundred folders of archival material.[44] The investigation offered here, though, will have to contend itself with proposing preliminary conclu-

41 For a short overview of historic "neighborhood rules and regulations" dating from the seventeenth century onwards, as presented by one of Schunn's contemporaries, see Orend, "Siebenbürgisch-sächsische Nachbarschaftsordnungen."

42 Schunn, "Der Start zu einem Hohen Ziel," 48.

43 Schunn went so far as to argue that the historic neighborhoods had constituted organic and complex bureaucracies: "Like a living and independent organism, the neighborhoods provided all those many things that a centralized sophisticated city bureaucracy tends to in its own jurisdiction." Schunn, *Nachbarschaften*, 8–9.

44 The bulk of literature on the neighborhoods was written by Hans-Achim Schubert and dates back to the 1970s, while his key interests mostly lie with their pre-1933 models. See Schubert, *Nachbarschaft*. Schubert also published a chapter on "Wilhelm Schunn's 'Völkische' Nachbarschaften—Wunsch und Wirklichkeit" in König's 1994 edited volume that offered a clearly enthusiastic overview of Schunn's achievements without discussing eugenics, save a short mention of the honorary gifts. That said, probably the most interesting article on the neighborhoods is offered by the sociologist Georg Weber's three-pronged analysis of changes to their "form and function" in pre-1939 Transylvania, post-1953 Germany, and 1980s Romania. While preoccupied with the church's rural neighborhoods rather than with Schunn's urban "national" ones, he fully grasps their utility as coercive tools and their complementary use of symbolic imagery. See Weber, "Zum Struktur- und Funktionswandel der siebenbürgisch-sächsischen Nachbarschaft."

sions on the specifically eugenic and race-hygienic discourses as defined by the neighborhoods' ideological imperatives and methodological means, but does so with the ardent hope and expectation that this can be but the beginning of historiography's interest in the neighborhood system, which offers such a valauble prism through which to view the political trials and tribulations of the 1930s and the socioeconomic realities within which it operated.[45] This study looks at three overarching parts: the first investigates Schunn's reinvention and institutionalization of the Hermannstadt model; the second tackles the ways in which the model was exported throughout all the various German settlement areas with varying degrees of success; and the third describes how the neighborhoods were turned into local party "cells" for Andreas Schmidt's nascent NSDAP of the German National Community in Romania in 1940.

Form and function: The Hermannstadt model re-enchanted

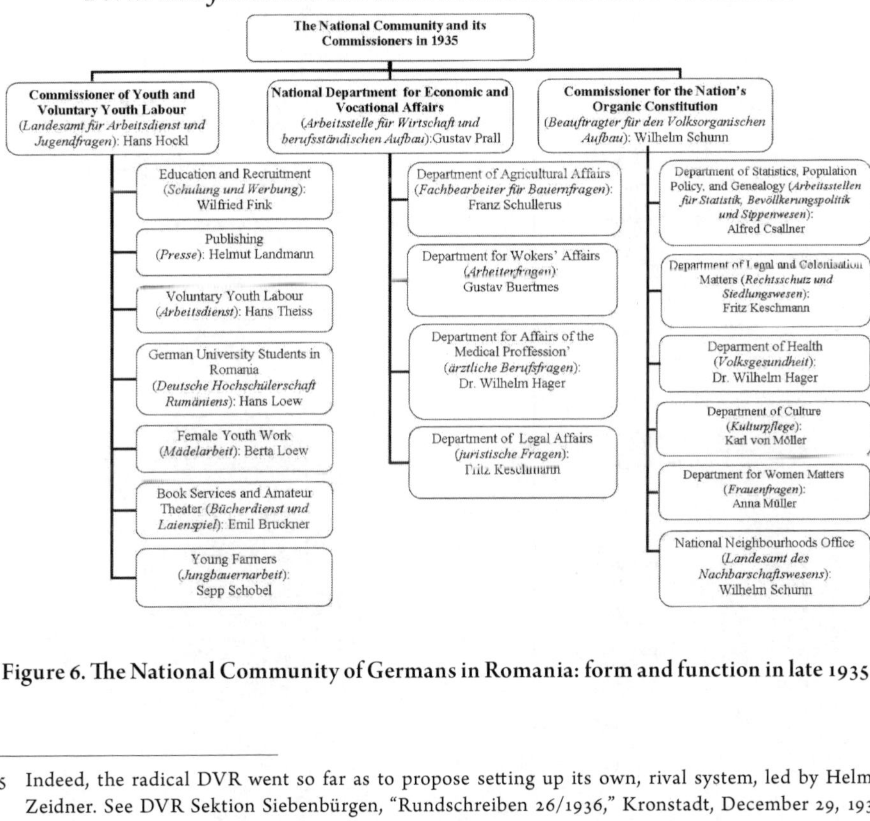

Figure 6. The National Community of Germans in Romania: form and function in late 1935.

45 Indeed, the radical DVR went so far as to propose setting up its own, rival system, led by Helmut Zeidner. See DVR Sektion Siebenbürgen, "Rundschreiben 26/1936," Kronstadt, December 29, 1936, CGR.D.XIII.A.230, pp. 1–2.

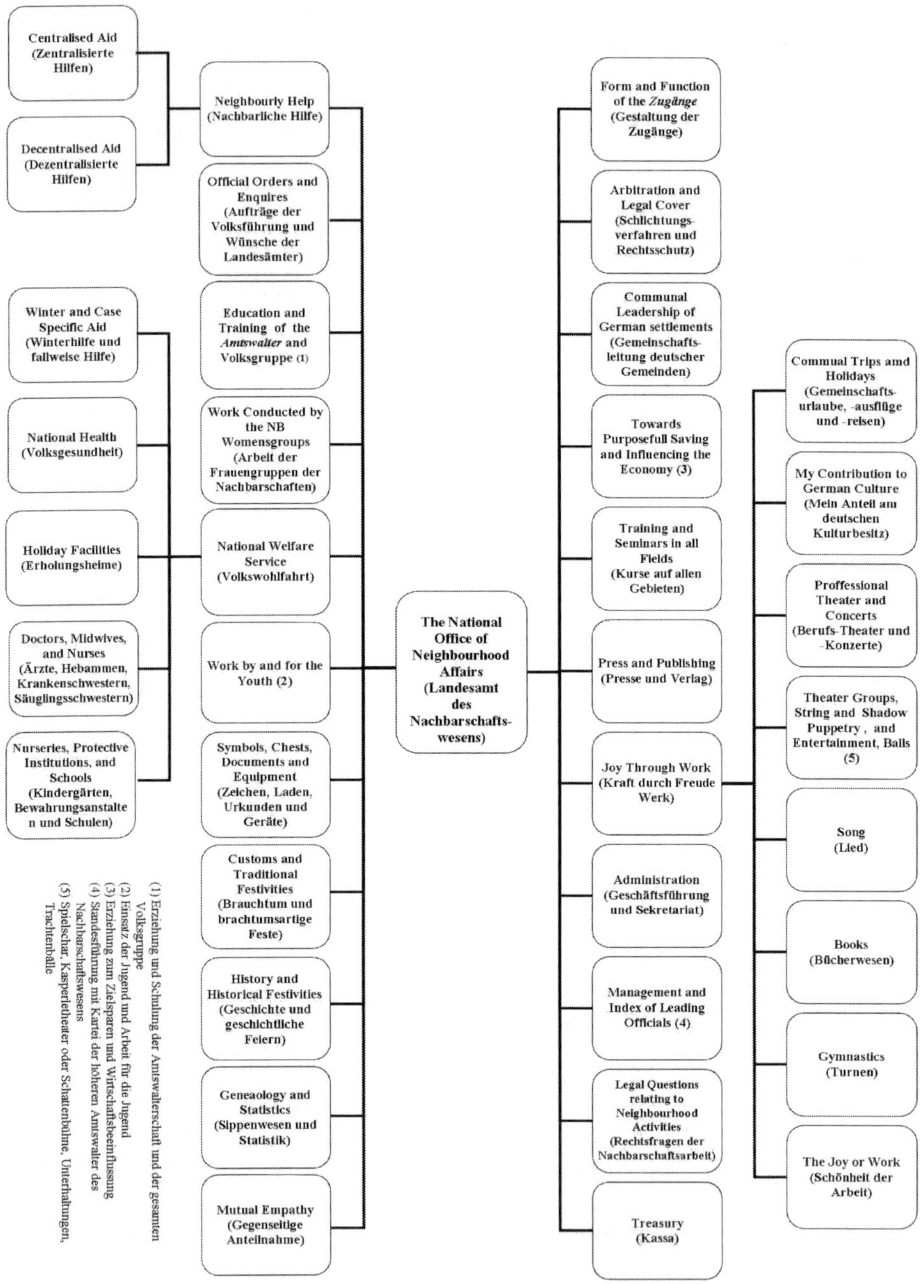

Figure 7. The Neighborhoods National Office and its twenty-two departments in 1939.

The neighborhoods, in their function as closed community support networks, handed fascist Saxon politics the singularly most potent tool with which to enforce policy decisions locally. Their provision of financial aid on the one hand, and ability to exclude undesirables from its ranks on the other, made them a formidable force. Schunn was also a close friend of Alfred Csallner's, and the neighborhood policy of awarding a honorary gift to celebrate the birth of a family's fourth or more child was deeply indebted to their collaboration. This "gift" came in the form of a one-off payment of a staggering 20,000 lei (all the more so considering that the overwhelming majority of recipients had a monthly income around the 2,000–3,000 lei mark) for the fourth, and a further 10,000 lei for each subsequent child. This money was conditional upon the submission and approval of a spending plan, and on a race-hygienic "hereditary fitness" exam (at least in principle) in the repeatedly stated bid to ensure that this incentive to found larger families only benefited families perceived as healthy and valuable. The terms and conditions framing the cash gifts hence amounted to a form of eugenic welfare. The gifts' underlying purpose, precisely due to the attached strings, hence also institutionalized a new, distinctly biological definition of Saxon identity, while threatening the exclusion of members that failed to live up to an imaginary racial and moral bar. In other words, the neighborhoods aptly demonstrate the significance and perceived utility of a eugenically defined vision of national regeneration.

Towards these goals, the neighborhoods offered the Self-Help three essential and distinct means: first, a virtually free bureaucratic network operated by hundreds of officials working on an honorary basis; second, a remarkably potent ideological tool with which to re-enchant public space and time, nurturing an "intimate" sense of communal destiny and duty; and third, a eugenically grounded welfare system. They also symbolized the regenerative project's attempt to fence Saxon society (and, by extension, those of the other German settlement groups after 1935) by virtue of ethnic exclusivity. The neighborhoods were hence an attempt at internal consolidation and regeneration in a perceived climate of perpetual victimization at the hands of an unreliable Romanian state, as illustrated, for instance, by Schunn: "We, the Germans of Romania, are sick of constantly losing in all domains. The constant struggle to merely defend, even if successful, leads to a spiritual degradation. We also have to build, as visibly and beautifully as possible, so that we may retain our confidence in our industriousness."[46]

Schunn believed to have found the ideal symbol of national self-reliance and resilience in these neighborhoods, historic institutions that enjoyed both an innate sense of legitimacy and mysticism, a relic of "old-Germanic communal life that has been lost to all other German tribes. This makes the Transylvanian neighborhoods all the more precious, like a holy tradition that we alone have preserved."[47] Indeed, Schunn argued that every "creative" (*schöpferisch*) nation had to design and live in accordance with rules and regulations reflecting its soul's characteristics, or else be ruled and ruined by that imposed by another. That said, it is hardly surprising to find Schunn advancing the ideologically mystified neighborhoods as the "living and independent organism" derived from the ancient Nordic *Markverfassung* (that for its part supposedly stretched as far back as the Stone Age).[48]

So, in what state did Schunn find these historic neighborhoods in 1933? While they had persisted piecemeal after being banned in 1891 (predominantly in rural areas, where they had been taken over by local church parishes), they had largely lost their form and function by the interwar period.[49] All the more reason to briefly note that, in a rather remarkable way, Schunn's neighborhoods also symbolized the Self-Help's ultimate triumph over Rudolf Brandsch and his Hermannstadt-based Bürgerabend, which had transformed them into local party branches in 1911. One of the few articles featured in the *Selbsthilfe* to foreshadow this irony, if before the Sachsentag, lamented that after the First World War "the neighborhoods appeared to lose their sense and purpose, which is why they were shut down, which is why they were neglected. They remained ownerless property [*Gut*], partially used for social events or party-political interests in cities and the countryside. That is how it was possible, for example, for the neighborhood in Hermannstadt to become a party matter for the Bürgerabend. [...] This cannot

46 Schunn, *Nachbarschaften*, 3.

47 Ibid., 6.

48 Schunn, "Die Nachbarschaften als geeignete Lebensform."

49 One of the few, but all the more interesting references to the neighborhoods system found in the *Selbsthilfe*'s pages is a 1928 article demanding they be awoken from their "twilight slumber" and restored as sentinels of healthy national life: "The neighborhood must be guarantor and protector of the individual family and, like it used to, must ensure that the nation's and family's fundamental pillars—discipline, morality, honor, and purity [*Artreinheit*]—are not disturbed." R., "Die Nachbarschaften und die Selbsthilfe," 1. See also Schunn, "Der Start zu einem Hohen Ziel," 47.

be tolerated."[50] And, clearly, it was not. Wilhelm Schunn outlined his vision for the new national neighborhoods' form and function in three key books that by virtue of including their various guidelines form the obvious point of departure for any serious discussion thereof.[51] But they are, of course, in many ways visions of want rather than feasible policies conceptually and practically, as illustrated by the omnipresent issue of how to deal with mixed marriages—that they largely omit—in a region the ethnic lines of which were far fuzzier than dogma demanded. In other words, and in addition to the host of problems bedeviling the neighborhoods' expansion beyond Hermannstadt, the analysis offered here is in many ways one of Schunn's ideal rather than of implemented reality.

The first set of provisional guidelines for the reconstituted Hermannstadt neighborhoods was legislated by the Saxon National Council's first curate session following the fifty-three delegates' return on January 22, 1934,[52] with a version applicable to all of Transylvania released in June 1935.[53] These were revisited to facilitate their export to all German settlement areas following the Self-Help's empowerment in the Association of Germans in Romania in 1936,[54] and a final set of guidelines was issued in July 1939.[55]

The neighborhoods were by design thoroughly suffused with a sense of re-enchanting public time and space, spawning a system of sacred geographies and symbols that legitimized their authority and grounded the Self-Help's project to reconnect with the historic in an alternative modernity. To use Arthur Moeller van den Bruck's expression, the neighborhoods were themselves a "*Wiederanknüpfung nach Vorwärts*."[56] A wonderful example of

50 Hager, "Die völkische Nachbarschaft."

51 In chronological order: the first, published in two editions in 1936 and 1937, outlined their overall structure and purpose; an instruction manual for neighborhood bureaucrats followed in 1939; and a propagandistic summary of the regenerative impact of the honorary gifts in Hermannstadt was published in 1940. See Schunn, *Nachbarschaften: Schulungsbüchlein*; Schunn and Pastior, *Die Ehrung des Kinderreichtums*. The fact that all of these litter German libraries makes it all the more perplexing that the neighborhoods in general, and their eugenic core in particular, have been overlooked so comprehensively.

52 "Die Satzungen der Hermannstädter völkischen Nachbarschaften. Beschlossen vom deutsch-sächsischen Volksrat für Siebenbürgen am 22 January 1934," NA Sibiu, CGR, D.XIII.A.205.

53 See Deutsch-sächsischer Volksrat für Siebenbürgen, ed., *Die Satzungen und die Nachbarliche Hilfe*.

54 See "Die Nachbarliche Hilfe: Verbindliche Weisungen an die deutschen Nachbarschaften in Rumänien; Beschlossen vom Vollzugsausschuss des Volksrates der Deutschen in Rumänien am 7 Januar 1936," in Schunn, *Nachbarschaften*, 87–101.

55 See "Organisationsrichtlinien des Nachbarschaftswesens," in Schunn, *Schulungsbüchlein*, 5–17; and "Die Nachbarliche Hilfe: Verbindende Weisungen an die deutschen Nachbarschaften in Rumänien," in ibid., 18–39.

56 See Griffin, *Modernism and Fascism*, 177–78.

this process is offered by Viktor Quandt, who having insisted that "the holy foundation of any culture fit for survival is the purity of its blood," explained how and why the new neighborhoods were ideally suited to the Self-Help's fascist worldview because they were "not only an ancient, mature [*gewachsene*] way of registering the entire national group, but it also became evident that its fundamental values, kept alive and strong through the centuries, overlapped with National Socialism's key demands to such a degree that it represented a delightful [*glückliche*] synthesis between ancient traditions and German renewal."[57]

As such, every neighborhood was conferred with its own sense of identity as defined by its territory, symbolic name and number, coat of arms, and otherwise sacred paraphernalia, such as the revered neighborhood chest, in order to link the old with the new, to infuse them with the aura of ancient and proper traditions reflecting a redefined sense of ethnic self.

So, for example, the Hermannstadt neighborhood covering the city's original settlement area and sitting on the site of its former church fortress was given the name "Under the Castle" ("Unter der Burg") to symbolize the location's historic significance. Keeping in mind that the Self-Help and eugenic discourses never tired of decrying the foreign "infiltration" of "Saxon" cities, this approach to naming the neighborhoods offered a remarkably illustrative means of creating a sense of communal destiny, a particular sense of blood and soil remodeled as an educational tool to suit the urban context. Apart from the names, Hermannstadt's historical development was also reflected by the numbers assigned to the particular neighborhoods, so "Under the Castle" became NB 1, with the subsequent thirty-eight numbered in the order in which their geographic locations had grown and been incorporated into Hermannstadt. This fusion of symbolic historic traditions with practical organizational and educational agendas is summarized by Schunn when he explains that:

> Today every child [...] knows that there [NB 1] the city's oldest settlement area lies, that it then spreads into the valley, the lower city, and that it—as the other neighborhoods' numbering shows—later spreads into the contemporary upper city that initially had been occupied by fortifications. The neigh-

57 Quandt, "Die Nachbarschaften der Deutschen Volksgruppe in Rumänien," NA Sibiu, CGR, D.XIII.A.238, p. 133.

borhoods with lower numbers are proud to have the longest history. That this form of numbering also offers its advantages when it comes to marches and the Neighborhood City Council [*Stadthannschaft*] internal administration—which today nobody would want to miss either—is happily admitted.[58]

Further synthesizing historic motifs with regenerative agendas, Schunn's neighborhoods introduced coats of arms designed by the LSS's heraldic expert Albert Arz von Straussenberg and his brother-in-law, the graphic designer Hermann Lani. These insignia were not only seminal to the process of identity re-formation, but equally useful for marches. Two further tools towards endowing the neighborhoods with a sense of historic belonging came in the form of the otherwise utterly superfluous "neighborhood chests" and the "wandering signs." The chests, the historic counterparts of which had held all paperwork relating to the particular neighborhood, were now assigned an essentially religious significance. They posed as relics in those instances where older versions survived, while new ones were ceremoniously "consecrated" (*Ladenweihe*). Their importance was elemental, and they were awarded a central role in all festivities and symbolic rites: "The chest has long become a symbol [*Sinnbild*] and this author [Schunn] has never met anyone who seeing such a chest was not filled with reverence [*Ehrfurcht*]."[59]

The "wandering sign," in its historic function as a means of spreading news and orders amongst neighbors, was equally superfluous. But perhaps not quite entirely so, as its use offered the neighborhoods a form of communication that bypassed the press and could hence operate in relative secrecy. More importantly, its key function was symbolic, in that it required one neighbor physically pass it on to the next, and hence encouraged active involvement in the process.[60]

Complementing these attempts to re-enchant lost historic institutions through signs and symbols, the neighborhoods were also endowed with a sense of sacred time, as defined by the rites underlying their various assemblies and rallies. The mostly monthly "officials' evenings" (*Amtswalterabend*) hence had the explicit purpose of exciting the enthusiasm of the neighbor-

58 Schunn, *Nachbarschaften*, 19.

59 Ibid., 29–30.

60 Ibid., 30.

hoods' foot soldiers, of enforcing a sense of sacred duty, communal destiny, and struggle "reflecting the new German man's spiritual views [*Haltung*]."

They represented the levelling political rallies Schunn believed had made "the neighborhoods a first class tool of spiritually [*seelisch*] influencing" the nation at large.[61] And so, the neighborhoods were inherently deeply dogmatic; despite all protests, they had a supra-political social quality removed from the chaos engulfing public life at the time. Unsurprisingly then, the "officials' evenings" were divided into an "official" and an "ideological" part, while at the local level "gatherings" (*Zusammenkünfte*) formed the individual neighborhood's social backbone, and were similarly divided into two parts: one "official" and one "casual." The neighborhood's educational purpose, though, demanded that the latter not descend into a "booze-up" (*Saufgelage*), by imposing (if not always enforcing) bans on alcohol and tobacco.[62]

The singularly most important event in the neighborhoods' calendar was the annual Day of Judgment (*Richttag*) that chronologically subsumed four distinct rites: the first, called *Blasi*, was a fun-festival for children, designed to reinforce their central role in national life; the second, the actual Day of Judgment, marked the neighborhoods' approach to public accountability and due process, where the male and female "elders" (*Nachbarhann/in*) reported their achievements, as did the treasurer, before the assembly proceeded to elect new elders. Following this annual progress report and election, third, a stringently structured communal dinner (*Kamderadschaftsmahl*) symbolized the community's unity, while "whoever is absent without a reason thereby shows that he despises [*verachtet*] the community and is not worthy thereof."[63] The fourth and final part of the Day of Judgment came in the form of a Fun Night (*Tolle Abend*), the merriment of which was thoroughly organized and prescribed by the elders, who had to approve whatever songs or plays made their way into the program.

On a final note, the "official" part of the Day of Judgment served to reinforce the sense of sacred symbols and communion above and beyond public accountability, through a number of rituals revolving around the increasingly sacralized "neighborhood chest." Indeed, the proceedings were opened by "the youth" ceremoniously carrying it into their midst, while chanting

61 Ibid., 31.
62 Ibid., 36.
63 Ibid., 38.

the following telling chorus:[64]

> *Caller:* You neighbors, you esteemed national brothers! / The chest comes, stand to welcome it!
> *The entire youth:* We are the youth!
> *The girls:* We are the girls.
> *The boys:* We are the boys.
> *The entire youth:* We bring the chest to the court!
> *Caller:* We are the young, you are the old! / Where were we all those hundreds of years ago? / Where were we all those thousands of years ago, / While our ancestors already sat to court?
> *The entire youth:* We were there, you were there, we were all there long ago, / We don't live to sixty, we don't live to eighty, / As long as our nation lives, so long shall we live.
> *Caller:* What is to be judged [*zu Gericht stehen*]?
> *The entire youth:* How long our nation lives, that is what stands trial. / We are the younger; we are only allowed to strive [*drängen*]. / You are the older, you must achieve it, / How long our nation lives stands trial.
> *Nachbarhann*: Place down the chest. / (pause until this is done and the chest opened) / Today we are the judges, / But time will once judge us. / You are the youth, you brought the chest, / The symbol of the past. / We thank you.
> *The entire youth* (slowly marching away, so that it has the time to repeat its final lines): How long our nation lives, that is what stands trial. / We are the younger; we are only allowed to strive / You are the older, you must achieve it. / How long our nation lives stands trial.

This chant underlined the perception of the nation as eternal, as perpetually renewed and regenerated by its youth that judges its elders but will itself, in time, be judged by its offspring—with the chest posing as history incarnate. But the underlying current of the pushing, striving youth bringing the chest to court is a key hallmark of not merely Saxon fascism, of course, but inherent to any political ideology that "believes" in the generational process of permanent renewal.

64 Ibid., 39–40.

Figure 8. The Hermannstadt Neighborhood Konradwiese: A life in pictures, 1936–39; "The Chest," Richttag, January 28, 1939.

Figure 9. The Hermannstadt Neighborhood Konradwiese: A life in pictures, 1936–39; "Communal Supper," Richttag, January 28, 1939.

Sacred money: Honorary gifts and the primacy of eugenics

Although the Self-Help had initially emerged as an urban ideology preoccupied with the urban poor, it quickly adopted the peasantry as the idealized source of the nation's hereditary and spiritual strength. Based on the familiar equation, the movement understood the frontlines in the struggle for *Lebensraum* to lay with the embattled rural communities that supplied the increasingly foreign Saxon cities with reinforcements through migration. The neighborhoods inadvertently assumed responsibility for their welfare. The honorary-gift model was its heart piece, designed to not only provide fiscal incentives to increase fertility rates, but tip the economic balance in favor of families deemed the most valuable hereditarily. It marked the institutionalization of a eugenic vision of regeneration deeply indebted to the "child enthusiast" Csallner, Schunn's main advisor regarding the rural implementation of this remarkable eugenic policy.[65]

All of the neighborhoods' social and eugenic programs were executed and managed by the department Neighborly Help (NH), under Viktor Quandt's stewardship (who was later succeeded by Siegmund's close friend Julius Ernst Gyurgyevich).[66] If, so to speak, the neighborhoods proper busied themselves with collectivization, ideological training, and political mobilization, the Neighborly Help was its welfare office. Although its two flagship policies focused on the celebration of life ("honorary gifts") and death ("contributions to funeral expenses"), it tended to a host of further socioeconomic policies on social health and hygiene (predominantly tended to by the women's groups);[67] communal aid in the case of disasters;

65 Ibid., 54.

66 Quandt actually wrote a rather interesting book about the Neighborly Help (even if it skirts around the issue of eugenics) in 1971, entitled *Neighborhood: Manageable Unities with Heart.* Apart from memories from the "good old days," over a third of the book is dedicated to describing the "new" neighborhoods he founded in Vienna in 1958. The rest of his book discusses the neighborhoods' historical evolution, merits as a form of social organization, and outlawing in the nineteenth century. Quandt proceeds to laud Schunn's "genius" in reviving them after 1933, describes the political turmoil that effectively stopped their spread beyond Hermannstadt, and laments their destruction by Schmidt. See Quandt, *Nachbarschaft, die überschaubare Einheit mit Herz.*

67 While the neighborhoods encouraged women's active engagement and participation, and every neighborhood had both a male and a female elder (*Nachbarhann/in*), the ideological insistence on traditional gender roles was uncanny. Of the eight goals that Schunn and his Commissioner for Women's Work Liane Klein set for the German Women's Group in Romania (Deutschen Frauenschaft Rumäniens), the first demanded "the mental and spiritual Betreuung and ideological training of the German women in the spirit of national renewal," while the second prioritized "the education and preparation of girls for

and, most importantly, offered "case-specific aid" (*Fallweise Hilfe*) that was overwhelmingly used towards covering medical costs.

By far the most important policy in scope and impact, the honorary gifts empowered the eugenic discourse through the provision of fiscal incentives to found the large hereditarily valuable families demanded by the eugenic and fascist movements. The neighborhoods hence succeeded in giving children the sanctified status that Csallner had demanded for so many years, and were omnipresent in celebrating them. Following a child's birth, the neighborhood ceremoniously handed the parents a "note of gratitude" to thank them for their new comrade, along with a pair of knitted shoes the child had to wear as a symbol of his or her appropriation by an "extended family" that demanded "he belongs to all of us and we want to take part of his care onto our shoulders."[68]

Although initially implemented in Hermannstadt, the honorary-gifts scheme was primarily aimed at peasants, and constituted the neighborhoods' solution to both a perceivably dwindling *Lebensraum* and the much decried loss of racial substance. Hence the stipulations that only provably valuable families were eligible for the staggering sum of 20,000 lei for the fourth, and 10,000 lei for each subsequent child born; and why a preapproved "spending plan" had to constitute an investment in largely inheritable goods, such as property or businesses. Striking two birds with one stone, the policy hoped to simultaneously boost desirable fertility while offering the means to sustain it: "The honorary gift is only paid out if the child's parents and siblings are hereditarily healthy and no moral [*sittlich*] concerns are raised (alcoholism, living out of wedlock). The parents submit a spending plan committed to the goal of improving the family's life [*Lebenslage*] through acquisitions of enduring value. The spending plan requires approval. The Neighborly Help has the right to itself determine the money's use."[69]

As these honorary gifts were paid by the community for (a service to) the community, they created their own sacred space in the form of a strictly structured ceremonial conferral thereof. At these occasions, the Neighbor-

the future job as wife and mother," and the fourth tasked them with "maintaining the health and racial worth of our national group." See Schunn, "Arbeitsbericht des Beauftragten für den Volksorganischen Aufbau," NA Sibiu. CGR.D.XIII.A.227, p. 10.

68 Schunn and Pastior, *Ehrung des Kinderreichtums*, 4. For a longer translated extract from this text, see Georgescu, "Transylvanian Saxon Eugenics."

69 Schunn, *Nachbarschaften*, 48.

Figure 10. Presentation of the Honorary Gift.

hood Council (*Altschaft*) or, ideally, the entire neighborhood, congregated to hear the City Council or Neighborly Help leader deliver a speech to the parents being honored, something, as Schunn stated, along the lines of "the German *Lebensraum* is once again defended collectively as it was in past centuries, the individual no longer stands alone, [...] the money presented here is holy money blessed by over twenty thousand comrades."[70]

Indeed, the very concept of "holy money" is rather intriguing, and frequently encountered. These religious connotations illustratively reflect the neighborhood's ideological synthesis of spiritual re-enchantment with hereditary regeneration. So, for example, the extensive set of preconditions demanded of potential candidates also spanned social issues (such as the parents being married), in addition to a race-hygienic evaluation:

> A medical certificate has to prove that neither the child's parents nor siblings have evidenced hereditary inferiority [*vererbbare Minderwertigkeit*]. If the respective child for whom the honorary gift is considered is him/herself

70 Ibid.

> inferior, the siblings though fully valuable [*vollwertig*], then the gift is nonetheless presented—but the parents lose their claim to any future child support, even if further children are fully valuable. The payment is not refused if despite the parents' hereditary diseases the prospect exists that their children will become desirable members of the national community.
> Alcoholism of one or both parents is considered a damaging disease for the offspring, and in such cases payment of child benefits can be refused.[71]

The hereditary evaluations were not only going to deliver verdicts on a family's perceived value, they were also a remarkable attempt to anthropometrically and medically examine the entire national community as it filtered through the system piecemeal. Indeed, Schunn never tired of insisting on these exams' fundamental importance to the Saxon experiment with national renewal, and are hence unsurprisingly omnipresent in the neighborhoods' literature and guidelines. More precisely, the exams fed three overarching goals: first, to act as means of social control and ideological suffusion; second, to evidence that the movement favored hereditary "quality" over sheer "quantity"; and, third, to illustrate their utility towards realizing one of Csallner's most ardent desires, the creation of a "hereditary archive."

The quantity vs. quality debate had largely been resolved by Csallner and Siegmund, both of whom clearly favored hereditary substance over numbers, and was firmly laid to rest with Schunn's insistence that "the recipients of our honorary gifts are under no circumstances fallen asocials, but in every case biologically desirable national comrades."[72] But the policy not only rewarded the worthy, it offered an "educational" and "practical" solution to the question of undesirable offspring: "The rule that the hereditarily diseased and hereditarily lazy [*erbuntüchtig*] are not allowed to receive the honorary gift has made wide sections of the masses aware that reproduction is undesirable in such families, and hence acts as the first practical measure towards combating undesirable offspring in our national group. The honors are proof of gratitude. A child not honored is an unwanted child."[73]

71 Ibid., 91–92.
72 Schunn and Pastior, *Ehrung des Kinderreichtums*, 43.
73 Ibid., 44.

But as "asocial behavior is also a ground for refusal,"[74] the honorary gifts unavoidably also offered the neighborhoods a powerful coercive tool to use in a eugenic nation (re)building project Schunn was keen to promote:

> This is the first practical and frequently even rather drastic defensive measure against unwanted offspring that, to my knowledge, a German national community has been able to introduce. Just imagine what it would mean for a family if the medical examiner rejects the honorary gift's conferral. Soon, of course, the entire neighborhood would know that further children in this family were to be sinful. So, if more children are born to it anyway, the bad father would have the entire neighborhood's public opinion against him and have to run the gauntlet between his outraged neighbors' gruff remarks.[75]

Schunn believed the honorary-gifts system would serve as a public educator (or at least as an enforcer), and ensure that only the nation's hereditarily best reproduce. It would appear the only thing he regretted was that it was impossible to simply evaluate the entire population beforehand ("we just have to accept as an unalterable fact that we are currently unable to examine the entire population in advance to the same standards"),[76] while the gradual aggregation of the race-hygienic data was destined to feed the "hereditary archive" that Csallner had hoped to place at the center of his quest to regenerate the Saxon nation.

On a final note, before turning to an evaluation of the tangible outcomes of the honorary-gifts policy, it is important to note that it has been almost impossible to find anything but circumstantial evidence that these flagship racial evaluations were actually carried out (while it must also be pointed out that a necessarily thorough investigation of the ca. five-hundred-folder-large archival treasure in the Sibiu National Archives might yet bring this proof to the fore). Apart from a handful of reports alluding to medical exams, but not discussing their methods or conclusions,[77] most of the paper-

74 Ibid., 9.

75 Ibid.

76 Ibid.

77 Here two examples of such rather circumstantial evidence. First, in July the family Lutsch was told the gift would only be conferred "under the condition that the medical examination demonstrates their biological full worthiness [*Vollwertigkeit*]" (CGR.D.XIII.B.460, p. 33). Second, when the family Müller applied for the gift in late 1937, the Neighborly Help faced a common dilemma: "The family Müller's medical examination has proved the family's health, which for its part would approve the gift. On the other hand though, it was realized that the children speak German badly, while they speak Hungarian and Ro-

work consulted during this research referred to the spending plans alone. So while there is an abundance of evidence that this centralized form of economic control was very much in place, the question of whether claimants did indeed undergo racial evaluations remains, for the time being at least, open. Nonetheless, the fundamental ideological importance that racial hygiene assumed should conclusively answer the questions of whether the Self-Help pursued a eugenic agenda in general, and that of whether it was (at least partially) able to translate it into a viable policy in particular.

From a more pragmatic point of view, these investments were also supposed to give legitimacy to the Self-Help's policies: "The otherwise moral impact even on those further afield should not be underestimated. A small house, an extension, a new business, one after the other created by the national community's means. Suddenly this national community, which a few years back was thought incapable of anything, has become unexpectedly important, and particularly so for the subsistence of our poorer comrades."[78]

So, what tangible benefits or outcomes did these honorary gifts produce, and how significant were they? Reflecting the central role they assumed in the eugenic vision of a racially stronger, purified, and closed community bound together financially and emotively through shared sacred spaces, Schunn actually wrote a book advertising the eugenic policy's achievements in 1940. Cowritten with Oskar Pastior, it had two central aims: first, to counter criticisms that these gifts were increasingly being "used" by the poor as a form of "economic planning," and second, to prove that they made a tangible improvement to their recipients' lives.

The question as to whether an increase in fertility that set in around 1935 was due to the gifts' concurrent introduction in Hermannstadt is, to Schunn, essentially secondary to the question of why its recipients had a fourth child in the first place. In other words, were children seen as tickets out of poverty? This was a crucial distinction in the public's mind, and had in fact been addressed, if only peripherally, by the policy guidelines insisting

manian all the better, and his declaration to raise his children in a German spirit has not convinced us enough to award the honorary gifts after six months" (CGR.D.XIII.B.460, p. 26). Instead, the Neighborly Help decided it would wait "until tangible results give us proof that the couple Müller are really raising their children as Germans" (Ibid., 27).

78 Schunn, *Nachbarschaften*, 51.

that the gifts' acceptance was mandatory for the rich and poor alike, in a bid to create the semblance of social equality. Nonetheless, it remained a question Schunn skirted around the fringes of by preferring to think of the gifts as an incentive for the wealthier sections of society (and hence all the more valuable according to Csallner) to have a fourth child to begin with.[79]

The evaluation of outcomes offered here focuses solely on Hermannstadt, the policy's initial point of departure, and the one for which consistent and detailed data is available. It is also divided into two segments, where the first reflects the numbers offered by Schunn and Pastior's book on the awards distributed by the neighborhood "Am Taubenbrunnen," and the second is a statistical overview of all of Hermannstadt's honorary gifts as evidenced in the donation lists (*Spendenausweise*) for the period April 1938–December 1940.

The Hermannstadt neighborhood "Am Taubenbrunnen" was number twenty-seven of the city's total thirty-nine, hinting at its geographic location on Hermannstadt's northern perimeter. Largely populated by migrant peasants whose birthrate outstripped the overall average, it was also almost twice the average size with 1,200 people. Accounting for the overwhelming bulk of his book, Schunn offers the statistics and complementary photographs of the honorary gifts awarded by "Am Taubenbrunnen" to a total of twenty-four families by 1940.

Reviewing the origins and socioeconomic makeup of the twenty-four families, it is striking that save four families that had one parent of urban origin, all were rather poor rural migrants. Given that the average monthly income was a meagre 2,565.25 lei, the prospect of receiving 20,000 for their fourth, and another 10,000 for every further child must have been enticing (and goes some way to explain why nine families received more than one gift). Similarly, the policy essentially fulfilled its primary aim (along with the Self-Help's founding motif of "clean and airy homes") by securing and expanding Saxon property ownership, given that twenty-seven of the thirty-four gifts were invested in property and housing (with two used to purchase farm animals, one unknown, and four used to repay debts).

A more varied picture is offered by the following statistics on the honorary gifts awarded in Hermannstadt as a whole between April 1938 and

79 Schunn and Pastior, *Ehrung des Kinderreichtums*, 6.

December 1940.[80] For the eight months in 1938 where data is available (namely, April, May, and July to December), a total of twenty-six honorary gifts were distributed in Hermannstadt, a number that only grew to twenty-nine in 1939 (while the data for June and September is missing). Of these twenty-nine gifts, sixteen went towards buying property or building thereon, and five were invested in the family's respective business (the money was in these instances commonly spent on a carriage and/or horse). Significantly, only two went towards covering existing debts. In 1940, sixty-three gifts were ceremoniously conferred, of which fourteen went towards clearing debts, thirty-nine were spent on property, and a further nine invested in businesses. While these numbers show that the gifts were primarily spent on expanding or improving the Saxon property portfolio, Viktor Quandt saw their most important achievement as a propagandistic one: "Due to their use and ceremonial conferral in all cases, as well as through the emphasis on honor, they have in defiance of all those know-it-alls already mostly achieved that which is this organization's goal, purpose—an economic blessing: The large family is today the neighborhood's pride and the national community's joy."[81]

The neighborhoods' remarkable ability to translate eugenic policies into practice was, of course, tied to their affordability. How could such a vast bureaucratic network, in addition to the cash gifts, be afforded by an ethnic minority lamenting its economic ruin? The system was actually rather straightforward, as the neighborhoods conducted two donation drives a month. One collected the regular monthly fees (which could be reduced or entirely waived depending on the individual's economic ability) towards covering the costs incurred by the bureaucratic apparatus—which, for its part, was comparatively cheap considering most positions were honorary. The second drive covered the Neighborly Help's expenses, which varied depending on the number of awards granted that month. Towards public accountability, the neighborhoods published a monthly newsletter, the *Donations List* (*Spendenausweis*) that offered statistical data on birthrates,

80 Compiled from NA Sibiu, CGR, D.XIII.B.354, pp. 1–50. In fact, Alfred Csallner was himself the recipient of a honorary gift in December 1938 (resident in NB 20), although the statistics sadly do not reveal what he used it for (Ibid, 1). This data is a fascinating source, and due to the depth and breadth of the details it includes, it certainly merits a more detailed, quantitative study.

81 Quandt, *Die Nachbarschaften der Deutschen Volksgruppe in Rumänien*, 134.

death rates, the families "honored," and the recipients of "case-specific aid" (including its purpose).[82] These are immeasurably valuable sources and must surely constitute one of the key sources to any investigation of Saxon economic trends in the latter half of the 1930s.

Towards ethnic exclusivity

The neighborhood's key purpose lay with furthering ethnic exclusivity and territorial re-homogenization. But as Csallner's numerous and increasingly radical rants against mixed marriages illustrate, there was an obvious disconnect between ideological programs and practical realities. And while the following is merely an introduction to the complex task of harmonizing the two, a thorough study of the neighborhoods archive may well offer the singularly most accurate reflection of why the Self-Help had placed them at the heart of its eugenic agenda, and why it was the most difficult and divisive policy to implement.

The exclusion of Jews from Schunn's neighborhoods was, of course, a crucial ideological and practical benchmark towards re-homogenizing the Saxon national body and *Lebensraum*. While the issue of anti-Semitism is too faceted to discuss at length here, Jews were clearly victimized by the movement for renewal's quest to build a eugenic fortress. Therefore, it comes as no surprise to find internal communications such as the following letter from the neighborhood's lawyer Dr. Julius Schorsten to the Hermannstadt Neighborhood City Council in May 1939: "Mr. Himmelfahrt is a member of the neighborhoods. He is supposedly a baptized Jew. I kindly ask you to research the matter and to ensure that he is excluded from the neighborhoods, should that be the case."[83] Investigating the case of a Mr. Zitron in August 1940, Schunn's letter to Viktor Fabritius actually quotes Csallner: "Having consulted the National Leader, I hereby inform you that a half-Jew, even if his behavior is exemplary, cannot even as an exception be considered a national comrade [*Volksgenosse*]."[84]

82 In a December 1939 letter to Emil Neugeboren, the author (presumably Schunn), characterized this approach to mutual investment and fiscal accountability as inherently superior to an average state's taxation system: "A state's taxation structure is, from our point of view, nothing but an organization opposing [*gegenüberstehen*] the mass of tax payers. The tax payers are not connected amongst each other, unless when they occasionally rot together [*zusammenrotten*] to destroy a tax law." NA Sibiu, CGR, D.XIII.A.278, p. 15.

83 NA Sibiu, CGR, D.XIII.A.250, p. 7.

84 NA Sibiu, CGR, D.XIII.A.237, p. 43.

The struggle towards ethnic exclusivity unavoidably found its key adversary in the much derided mixed marriage. What is more, the flurry of letters and memos on what to do about them aptly illustrates a crucial truth, namely that the neighborhoods did not have a clear definition of what being German actually meant above and beyond the rhetoric. Two illustrative case studies of how Schunn's organization tried to resolve the mixed-marriage conundrum are offered by the towns of Neumarkt and Strassbourg. In both cases, the neighborhoods had to maneuver around, rather than bulldoze through them, by introducing the Neighborly Help, even if with a considerably curtailed remit and lesser funds.

Finding a resolution for Neumarkt was particularly difficult, as its *Stadthann* Julius Haltrich was himself wed to a Hungarian wife. Similar to the above example of Mr. Zitron, Viktor Quandt subsequently asked Schunn (in October 1939) for advice on what to do with a man who had sinned against his nation but otherwise delivered exemplary work. All the more so as Quandt was "embarrassed" to have sent Haltrich a letter outlining a rather unequivocal approach to mixed marriages:

> Mixed marriages. Their admission is ruled out as a matter of principle, even the German part of such a marriage is unwanted. If, though, the German part is absolutely dominant and the family is completely German, then the foreign sounding name cannot be an obstacle to admission. But those mixed marriages where they do not think or speak German in or outside of the house are strictly declined. The experiences gathered over the past five years in Hermannstadt have led us to assume a very strict view on this matter, one that may often appear tough and unfair. But I kindly ask you to be rather too tough than too lenient in this regard, it is in any case the smaller of two evils.[85]

While this quote already hints at a certain conceptual fuzziness, the degree to which the neighborhoods were forced to sacrifice ideological imperatives in favor of political pragmatism (and, quite simply, reality) is clearly evidenced by the example of Strassbourg. Corresponding with its *Stadthann* Martin Riffelt, also in October 1939, Quandt's approach is characteristic: to deny membership to "new" mixed marriages, while accepting "old" ones, and to focus on providing help towards funeral expenses rather

85 NA Sibiu, CGR, D.XIII.A.278, p. 106.

than introduce the honorary-gifts model.[86] In fact, Strassbourg had already been branded a "lost post" after Riffelt reported only four marriages were entirely German, and even voiced his opposition to the policy's potential impact on community relations:

> But we have splendid and reliable German men with a Hungarian wife, and in these houses they speak German-Hungarian—and the children should be evaluated that way, too. But we all come together as Germans and one brings his Hungarian wife along and speaks German and Hungarian as well as one can. The neighborhood is like the church congregation. If the neighborhoods now introduce a strict separation, this will be projected unto the church congregation, and discord will spread. The German man cannot separate himself from wife and children because of the neighborhood. So, for example, our curator has German, Hungarian, and Romanian sons-in-law, and when all meet, they communicate well in all three languages! It is obviously not an ideal scenario. We cannot condone this situation either. But I am the congregation's priest, and it is what it is.[87]

Father Riffelt's remarkably frank assessment of the interethnic realities in Strassbourg, in spite of the Self-Help's and Csallner's increasingly radical demonization of mixed marriages, concludes with the equally frank demand to either let all "German-minded" (*deutschgesinnt*) join the neighborhoods or dissolve the local Neighborly Help, declaring that he would otherwise resign his post as *Stadthann*. The eugenic fortress's ideological mortar was in the end unable to reconstitute an ethnic exclusively eroded by centuries of intermarriage and cohabitation.

The neighborhoods' expansion in and beyond Transylvania

Given the overall success of the honorary-gifts policy in Hermannstadt, in union with the neighborhoods' ability to efficiently organize and influence local affairs, how successful was Schunn at exporting his Hermannstadt model in Transylvanian, let alone the other German settlement areas that had no previous experience of them? In short, he was not par-

86 NA Sibiu, CGR, D.XIII.A.282, p. 3.
87 Ibid., 8.

ticularly successful, especially in rural areas (where the neighborhoods largely remained church institutions) and outside of Transylvania (where the political battle lines frequently undermined his work). In addition, the system required a significant bureaucratic apparatus to function locally in the first place, and its members' willingness to pay for another's medical bills and fertility on the basis that they themselves might also benefit in due course. Neither of these two prerequisites was in high supply during the turmoil of the 1930s.

Nonetheless, the neighborhoods were, by comparison, to spread rather quickly and efficiently in Transylvania, particularly so once Schunn became the leader of the church's rural neighborhoods in March 1939.[88] Remaining with the region's cities, Mediasch was the first to adopt the Hermannstadt model, and began paying out its first honorary gifts in mid-1936,[89] while their expansion to Schässburg descended into a bitter political battle and the *Stadthann*'s resignation. Similarly, Schunn's experiences with infiltrating Bistritz were disappointing, to say the least. Political turmoil resulted in more resignations, and Schunn eventually resigned himself to the realization that Bistritz "is the only larger German city in Transylvania where it seems unlikely enough that willingness to assume work will be found."[90] However, in a letter grudgingly accepting Georg Müller's resignation as Bistritz's *Stadthann* in October 1939, Schunn was considerably less eloquent: "I am quite desperate about Bistritz. I hesitate to say what hangs from my lips. I am filled with such an anger mixed with disgust that I swing between the two extremes of either wishing you may all be completely devoured by the Jews, or the extreme of not letting go until Bistritz has a decent neighborhood life and all the whiny gnomes [*jämmerliche Wichte*] have gotten a proper kicking."[91] It is curious to note that one of the obstacles hindering the

88 In a memo addressed to all of the National Community's departments, Schunn explained that this neighborhoods' merger came about in a bid to prevent rivalries and frictions as the national neighborhoods began expanding into rural communities. See NA Sibiu, CGR, D.XIII.A.278, p. 291. While Schunn continued to underscore that this union "clearly, and without a fight, increased the German renewal's influence, which is extraordinary," the church's memo announcing this landmark shift in jurisdictions was rather more subdued when it simply stated the move would "revive the neighborhoods through the energies of German renewal, while preventing the disruption [*Erschütterung*] of the church's neighborhood work." See Z.K. 560/1939, published in Kirchliche Blätter 15 [1939].

89 Schunn, *Nachbarschaften*, 69.

90 Schunn, "Der Start zu einem Hohen Ziel," 51–52.

91 NA Sibiu, CGR, D.XIII.A.278, pp. 99–100. See also another of Schunn's responses to a letter of resignation, in this case that of Dr. Albert Kräutner in May 1939, in ibid., 249.

neighborhoods' expansion in Bistritz was in fact competition. More specifically, the local "funeral clubs" (*Leichenvereine*) offered the same cooperative assistance towards funeral expenses provided by the Neighborly Help—and the two had to compete for members.[92]

If the cities posed largely political and organizational challenges, the neighborhoods' expansion into smaller communities offered a far more faceted picture of the successes and shortfalls that imposing ethnic exclusivity entailed. While Schunn was delighted to note their rapid establishment in some essentially non-Saxon cities—or the "Saxon diaspora," as he termed them—such as Neumarkt am Mieresch (in April 1936), where he claimed they had revived Saxon national awareness (and commitment) in the first place, the picture becomes far more hazy with places such as Fogarasch. Here, the Saxon "colony" subsumed some 700 individuals, of whom 600 were organized in neighborhoods. But as they were only open to ethnic Saxons/Germans, they created a real conundrum for cities which exhibited a relatively high degree of intermarriage: "A one-hundred-percent inclusion of all national comrades will hardly be possible due to the numerous mixed marriages and the semi-Magyarization of certain circles, nor is it desirable."[93] This reality, though, severely impacted the principle function of the eugenic population policy, and resulted in the piecemeal introduction of watered-down policies, in order to avoid becoming counterproductive race-hygienically.

When it came to introducing the Saxon neighborhood model in the Banat, Schunn's primary aim was to use Temeschburg as the springboard from which to expand into the surrounding countryside and beyond. Justifying the city's central role in this overall plan, Schunn insisted on the importance of winning the country's "largest German city" (with a German population of about 35,000), where the "Marxist workers" were in dire need of a social network to "make them more consciously German."[94] Work towards these goals began with the first official meeting convened by the new *Stadthann*, Dr. Josef Riess, in Februray 1936, and the city eventually subsumed forty neighborhoods, leading Schunn to conclude that: "The work's progress resembled a victorious march [*Siegeszug*] that far outstripped our expectations. We were especially delighted by the surprise of finding that

92 Ibid., 174–76.
93 Schunn, *Nachbarschaften*, 70–71.
94 Schunn, "Der Start zu einem Hohen Ziel," 54.

the Marxist workers did not have to be arduously won, but that they immediately joined in large numbers and themselves became the work's key drivers [*Träger*]."[95]

As for the Old Kingdom, there were few territorially coherent pockets of Germans other than the 20,000-strong congregation in Bucharest, which for its part was astonishingly well organized. Under Michael Deppner's direction, sixteen neighborhoods were created, complete with their own monthly newsletter (it is, however, impossible to reach any substantive conclusions about their activities, as only a few disjointed samples thereof are currently available).[96] On a final note, the neighborhoods also expanded into the Bukovina, but had little or no tangible impact (partially because their honorary gifts were not generous enough to be of "significant economic relevance").[97]

95 Schunn, *Nachbarschaften*, 73.
96 Schunn, "Der Start zu einem Hohen Ziel," 56; Schunn, *Nachbarschaften*, 75.
97 Schunn, "Der Start zu einem Hohen Ziel," 53–54.

CHAPTER VI

1940 AND EVERYTHING AFTER

The year 1940 was a cataclysmic one, ending Saxon eugenics' second (or interwar) period under the aegis of Siegmund, Csallner, and Schunn, and ushering in a third one, defined by the German minorities' *Gleichschaltung* with the Third Reich's needs and wants. This watershed, very appropriately, came in the form of Andreas Schmidt—Gottlieb Berger's son-in-law—appointed as National Leader of all of Romania's Germans in September 1940 by the head of the Main Welfare Office for Ethnic Germans (Hauptamt Volksdeutsche Mittelstelle, or VoMi), Walter Lorenz. Schmidt decisively changed the framework for Saxon eugenics, if not necessarily the core principles, with his creation of the NSDAP of the German National Community in Romania (NSDAP der deutschen Volksgruppe in Rumänien) in Mediasch, on November 9, 1940,[1] and offers a natural end date for this study, with the last vestiges of an independent movement drowned out by the wholesale import of German models along with a highly centralized bureaucracy that moved from Hermannstadt to Kronstadt.

The sociopolitical and geopolitical storm gathering in the wake of Schmidt's return to Transylvania has been researched and published on widely, witnessing Marshall Antonescu's empowerment in Bucharest; the arrival of German troops and the *Volkgsruppe*'s recognition as a juridical person that entailed significant autonomy; the Second Viennese Accord that split northern and southern Transylvania and awarded the former to an ever-irredentist Hungary; the creation of the Gau Bergland spanning the northern Banat and south-western parts of Siebenbürgen (Gauleiter being

1 See Schmidt, "Gründung der Nationalsozialistischen Arbeiterpartei der Deutschen Volksgruppe in Rumänien."

Hans Ewald Frauenhoffer); the resettlement of Germans from the Bukovina, Bessarabia, and Dobrudja; as well as the wartime effort that enlisted so heavily amongst the German minorities and proved so fateful.

Andreas Schmidt's descent on Saxon society and politics in 1940 also marked a generational change towards younger ideologues predominantly trained in Nazi Germany, resulting in the marginalization and eventual substitution of almost all of the eugenic fortress's interwar architects and their respective organizations. Some did, of course, slot into the institutions established by the new regime, such as Eckhard Hügel, who came to hold almost every office tasking with race-hygienic research and policy, Alfred Hermann, who became the head of the Further Education Department (Hauptabteilungsleiter für höheres Schulwesen), or Wilhelm Hager, who oversaw the school-nurses resort in Siegmund's Welfare Office and later became director of the Main Office for National Health (Hauptamt für Volksgesundheit).

Amongst the plethora of NS-German templates grafted upon Romania's German minorities by founding or adapting existing structures are, for example, the party political structure of the NSDAP, the Winter Aid Program (Winter Hilfswerk), the NS-Welfare (NS-Volkswohlfahrt), "Strength for Joy" ("Kraft durch Freude"), or the German Troupe (Deutsche Mannschaft) youth organization, and its Einsatz Staffel as an equivalent of the SS.[2]

Schmidt's *Volksgruppe* hence also entailed a fundamental redefinition of the form and function of Schunn's neighborhoods—one he was anything but pleased with. Schunn summarized the key aspects of his dislike for the new regime in a letter to Schmidt bitterly lamenting the neighborhood's "destruction" through the imposition of German (rather than Saxon) organizational models.

Considering the debate surrounding the issue of whether Saxon fascism had simply imported German National Socialism rather than offered an indigenous fascist ideology, Schunn's repeated insistence that Saxon society needed to be governed by organic Saxon structures merits a closer look. Schunn was absolutely adamant that neither the form nor function of the German NSDAP was appropriate to the Saxon context: "I think: the form

2 For a short overview of early organizational changes, see the short pamphlet by Schmidt's propaganda chief Walter May: May, *Ein Jahr Partei.*

of the party's work in the Reich was developed to pursue the political goals of one party against another, and did so—which one may not forget—in the Reich! Our national organization, though, is not tasked with the purpose of a political battle it cannot accommodate, but has to create a framework for our entire national community's national life, as it evolved over centuries."[3]

Schunn deeply resented the neighborhoods' appropriation by Schmidt's party that retooled them into its organizational hierarchy of "blocks" and "cells,"[4] and replaced the Neighborly Help and its honorary gifts with a social welfare office based on the German NSV model by Wilhelm Schiel.

More to the point still, and reflecting the neighborhood's attempts to both impose the Self-Help's *Führerprinzip* and keep the neighborhoods as semiautonomous clusters (that could, essentially, survive if the central office were closed), Schunn's letter also discussed the question of coercive means:

> But the methods, too, must be different in our case. The party in the Reich developed a vertical command structure from the leader to the led. I don't want to deny that this can be useful when actions need to be carried through quickly. We Saxons and Schwabs, though, are not concerned with carrying out fast actions, but our historic and eternally same concern lies with taking care that our national comrades remain attached to each other [*anhänglich*], that they stay together and enjoy each other's company, and so nourish and strengthen the horizontal! relationship between national brothers.[5]

The year 1940 was hence to mark the beginning of the end of Schunn's career, and he reemerged from the chaos of 1944, like Csallner, a priest. In October 1951, Schunn became the village priest in Schal (near Mediasch), and following a 1958 government ban stopping his attempts to reconstitute the neighborhoods,[6] Schunn could simply be "a Christian amongst Christians" in the community he served unto his death aged seventy-eight, on November 3, 1966.[7]

3 ZAEV, Bestand Schunn, no. 6, p. 1.
4 Quandt, *Nachbarschaft*, 21; Schubert, "Wilhelm Schunn's 'Völkische' Nachbarschaften," 177–78.
5 ZAEV, Bestand Schunn, no. 6, p. 1.
6 Schubert, "Wilhelm Schunn's 'Völkische' Nachbarschaften," 180.
7 Quandt, "[Obituary for Wilhelm Schunn] Der Wiedererwecker der nachbarlichen Hilfe." In fact, Schunn had been elected the "city preacher" of inner Kronstadt in 1957, a post he had to turn down, though, due to illness. ZAEV, Bestand Schunn, no. 3.

As for Csallner, following the LSS's 1940 expansion into the Institute for Statistics and Population Policy of the German National Community in Romania (ISB), one of his final assignments was to design and macro-manage the meticulous census of all German settlements in Romania carried out in November 1940 (in collaboration with Schunn's neighborhoods). The census served a dual agenda: to both furnish Andreas Schmidt's newly created *Volksgruppe* with a complete inventory of its human capital (in terms of population size and economic strength in relation to the supply and demand of labor), as well as lay the statistical groundwork that underpinned the VoMi-led resettlement of Germans from Bukovina, Bessarabia, and Dobrudja shortly thereafter.[8]

Csallner published the survey's results with remarkable speed that same year, under the title *The German Settlements in Romania as Seen in the Census of November 3, 1940.*[9] Regrettably, it offered no substantive analysis of the alphabetical list of all "German" towns and villages surveyed in terms of their German population's size in comparison to the Romanian 1930 census. Its most interesting feature was the appended, color-coded map by Günther Herbert, visualizing the geographic distribution of all villages and towns inhabited by at least 100 Germans (which offers a valuable comparison to Günther's map of Hermannstadt designed with the intention to illustrate precisely how many entirely "foreign" towns had come to besiege it).

That Csallner found himself robbed of the opportunity to digest the data in more detail was, at least partially, the result of the abyss that had begun to widen ever further between him and Andreas Schmidt. As Csallner was to repeatedly complain, Schmidt sent the questionnaires to the VoMi in Berlin to have them evaluated with the help of a Hollerith machine, despite Csallner's insistence they were not suited for it.[10]

While it is difficult to discern exactly why Csallner and Schmidt fell out, even if their different views on how to analyze the census data might well have been the trigger, their disagreement's escalation is aptly described by Wilhelm Klein: "One day, a *Reichsdeutscher* pharmacist by the name of

8 A point avidly made in Glass, "Wer ist ein Deutscher?," 26.

9 Csallner, *Die deutschen Siedlungen in Rumänien.*

10 See, for example, Csallner, *Zur wirtschaftlichen und sozialen Lage,* 176. What seems to have annoyed Csallner even further was that Schmidt had tasked "eine Berlinerin" (in whom he clearly had little faith) and had even forbidden her to talk to Csallner. See Csallner, "Die Rumäniendeutschen Bestandsaufnahmen," 218.

Schmierer came and informed him [Csallner] that the purpose of his visit was to investigate possible complaints against the national leadership. And when [Al]Fred trustingly told of his troubles, he was asked to put them all down in a memorandum—*relata refero*; shortly after complying with this wish, he was kicked out of office because Mr. Schmierer had nothing more urgent to do than place it in the hands of Andreas Schmidt."[11]

Although it has been impossible to find this memorandum, and considering that Csallner abstained from expounding its contents in later publications, the order of events that followed was rather straightforward. Schmidt dispatched Csallner to work with the VoMi in Berlin before dismantling the ISB and swaths of Csallner's work along with it in 1943.[12] Upon his return, Csallner found himself battling litigation in the NSDAP der VgrDR party courts, through which Schmidt had him declared "unworthy and incapable" of holding party office.[13] Having effectively ended Csallner's career, Schmidt also prohibited the Schools Office (Schulamt) from hiring him as a teacher in Hermannstadt, Kronstadt, Temeschburg, or even Bucharest, trying to sideline him with a high school job in the small Banat town of Reschitza instead—a post Csallner categorically refused to take.[14]

Yet, in spite of the turmoil and running battles with Schmidt, Csallner claims to have received two rather generous job offers. The first came in 1943, when the Südostinstitut in Munich elected him into their Research Advisory Council (Forschungsrat) and Friedrich Burgdörfer proposed Csallner run the Romanian branch of the International Organization for Population Science he had planned to found. The second, which would be particularly interesting if future research could substantiate it, was supposedly floated in early 1944, when Csallner was approached with the offer of applying his LSS template towards founding a new, larger institute in Prague, complete

11 Klein, "... lies Kirchenteppiche pfänden."

12 Csallner, "Die rumäniendeutsche Bestandsaufnahmen," 218: "But the relationship between the National Leader and myself soon worsened [...] so that he sent me to Berlin with a phony mission [*Scheinauftrag*], where I was supposed to be detained, and was in fact detained for four months. During my absence, he dissolved my institute with its many specialists and well-trained staff. Along with much more, the census material was lost, made useless, was destroyed." See also Wagner, *Laudatio auf Alfred Csallner*, 1–2.

13 Csallner, *Meine wissenschaftlichen Arbeiten* [1975], 7: "But tensions and differences between the National Leader Andreas Schmidt and me arose. He dissolved my institute and had the supreme party court declare me incapable and unworthy. I continued my fight against Schmidt until Romania's swing [*Umschwenken*] to Russia on August 23, 1944, and its repercussions ended that as well. Now I was locked up in a concentration camp."

14 Wagner, *Laudatio auf Alfred Csallner*, 1–2. See also Wa[gner], "Alfred Csallner 95 Jahre alt," 11.

with a lecturer post at the Karl University. Considering what followed the *Volksgruppe*'s collapse in August 1944, it is little surprising to find Csallner reflecting that "it wouldn't make much sense to try and envision how much differently everything would have been had I accepted."[15]

Although Csallner's age spared him from being deported to Russia, he repeatedly insisted that his daughter Trude's death (who was also one of his closest coworkers) had been the consequence of her "painful destruction" by forced labor in a Russian work camp.[16] It was also for Trude that Csallner initially composed the *Memories of Bistritz* (that were expanded into his *Rottenholz and the Rottenholz Grandparents*) while interred in Târgu Jiu and Turnu Măgurele between late 1944 and fall 1945. Following his release, Csallner found employment as the priest of Roseln,[17] Durles, and Petersdorf (near Marktschelken), and spent another eight months in jail before retiring aged seventy, in 1965.[18] Following his move back to Hermannstadt, Csallner was arrested again, sentenced to five years, but released after seventeen months imprisonment.[19]

In 1974, Csallner immigrated to Western Germany in the company of his wife and youngest son, where his legacy was quite unexpectedly revived, redeemed even. Without necessarily wanting to engage with the criticism—at times justified, at times polemical—levelled at the Transylvanian Saxons' postwar umbrella organization the Landsmannschaft der Siebenbürger Sachsen, it appears at least interesting that they awarded Csallner their Cultural Award (*Kulturpreis*) in 1982. The justification offered by the awarding body is rather revealing:

> The Landsmannschaft of the Transylvanian Saxons in Germany and Austria awards the Transylvanian Saxon Cultural Award for 1982 to the retired statistician and population scientist Father Alfred Csallner.

15 Csallner, *Meine wissenschaftlichen Arbeiten* [1975], 26–27.

16 Csallner, *Zur Verleihung des Kulturpreises*, 7. See also Csallner, *Rottenholz*, 6.

17 Csallner struggled to find peace in Roseln, and his short stay there might well be explained by the following, rather entertaining account offered by Wilhelm Klein: "After the war, I found [Al]Fred as the priest of Roseln. He had difficult times there too. Amongst his most ardent complaints was, for example, the following: The Roslers pitied the widow of the so tragically deceased father Lutsch, let her live in the vicarage, and frequently brought her numerous gifts that Fred thought ought to be his. He frequently quarreled with the widow and, to be objective, I must say that Mrs. Lutsch—by the way, a *Reichsdeutsche*—was a difficult woman to digest, and one to cleverly use her advantages." Klein, ". . . lies Kirchenteppiche pfänden."

18 Wa[gner], "Alfred Csallner 95 Jahre alt," 11.

19 Csallner, "'Es ist fast nichts übrig geblieben,'" 2.

> Alfred Csallner has made discoveries in his wide array of statistical and demographic studies. On his initiative, the Germans of Romania founded the National Office for Statistics, Population Policy, and Genealogy in 1935, which he directed unto its dissolution in 1941. To the Transylvanian Saxons, Alfred Csallner's research was of existential value. His only partially survived works are of high documentary value. He has made his findings accessible to the wider population, and attempted to encourage them to live in accordance with them.
>
> Alfred Csallner has rendered a service to the Transylvanian Saxons. In recognition and gratitude for his lifelong work, he is awarded the Transylvanian Saxon Cultural Award.[20]

Although the text reflects the empirical value of the vast databases that Csallner's LSS amassed, it skirts around the ideology that fed them, as well as the increasingly radical eugenic population policies they served.

Perhaps one of the most fundamental, and certainly consequential structural reforms introduced by Schmidt related to the Saxon Protestant church and its complete subordination to the *Volksgruppe*. The church's loss of competences such as the school system to the Schools Office (*Schulamt*) and other parts of rapidly expanding political structures concluded the turbulent relationship of the 1930s. It also accelerated dramatically with Bishop Viktor Glondys's resignation and immediate replacement with Wilhelm Staedel in February 1941.

The church's subordination marked a turning point for the Welfare Committee, too. In January 1939, Georg Müller took over from Walter Klein, who had acted as a caretaker director after Siegmund's death, and was quick to outline his three core priorities as "1. healthcare in its narrowest [*engen*] sense, 2. population policy, 3. racial hygiene [*Rassenpflege*]."[21] Müller, coming from Noesen, had to relinquish his post at the end of the year, with the partition of Transylvania.[22] As for the *National Health* exhibit, visitor numbers were unrewardingly low in 1937 and 1938, with 113 members of the public and 277 as part of tours for the former, and a drop to 96 and 221,

20 "Kulturpreis für Heinrich Bredt und Alfred Csallner."

21 Müller, "Die Aufgaben unserer Kirche, Teil I," 5. For Müller's views on Saxon racial composition, a glowing endorsement of the Third Reich's sterilization law, and thoughts on whether all church employees and settlers ought to be required to submit reports on their hereditary fitness, see Müller, "Die Aufgaben unserer Kirche, Teil II."

22 See Müller, "Zum Abschied."

respectively, for the latter.[23] In 1941, apparently at least in part due to the onset of war, numbers dwindled further, and the exhibit's director Julius Erst Gyurgyevich's report laments that only 89 regular visitors (of which 55 were two school groups) had come, while there had been 51 special tours with 433 people, as well as two prominent German visitors, Prof. Hans Holfelder, who was conducting X-ray health assessments, and the German Foreign Office's Fritz Theil.[24] The exhibit itself, after Gyurgyevich's initial reservations about losing independence, was absorbed into the *Volksgruppe*'s Propaganda Office in 1942.[25]

Another illustrative example is the tuberculosis sanatorium the church created in Hermannstadt, which was governed by a board of trustees including Bishop Glondys as chair, Schunn as his deputy, as well as Georg Müller and Helmut Wolff. When Müller had to resign in 1940, Schunn quickly nominated Dr. Gerhard Dietrich, director of the NS-Welfare (NS-Volkswohlfahrt).[26] In February 1941, the Office of Health Services' Matthias writes to the church asking for the sanatorium to be handed over, pointing out that it had been created with the neighborhoods' help and that his office would cover the 182,000 lei deficit. When the church asks the trustees for their views on the matter, Dietrich rather unsurprisingly responds to say that it should hand the sanatorium over, along with any remaining funds from donations:

> The medical management and welfare are planned and executed by the Office of Health Services (*Amt für Gesundheitswesen*). Its competences hence lie quite beyond the church's sphere of influence. [. . .] The national church can claim for itself the indisputable merit that at a time when welfare and national health provisions were in dire straits, as a true national church, it took these matters into its portfolio and dealt with them as well as it could. It will all the happier greet their professional and sufficient supervision through a specialized office, and supply all the resources it has collected for a task that it only took on out of necessity.

23 See Müller, "Das Landeskirchliche Fürsorgeausstellung 'Volksgesundheit.'"

24 ZAEV, LK 103.101 (1942), 31.

25 For the correspondence and negotiations regarding this, see ZAEV, LK 103.99 (1940), 345; and ZAEV, LK 103.101 (1942), 31. Also see Gyurgyevich's 1939 autobiography, including a discussion of his work for the Order of Good Templers, Csaki's *Kulturamt*, and his early involvement with the Self-Help, in ZAEV, LK 103.98 (1939), 53.

26 ZAEV, LK 103.99 (1940), 707.

> The Office of Health Services alone is responsible for health-political, medical management, planning, welfare, etc., and because it has to plot its field of activity rather widely, will keep a watchful eye to see that no parallel work is conducted.
>
> To go one's own way here would not be appropriate, as it would be mean to strive towards the final goal of battling national plagues with only half the resources.[27]

That Dietrich came to write this report through Schunn is indicative of a process whereby political actors gradually ascended to key roles in church structures, and it is little surprising that the Landeskonsistorium replied with its decision to cede the sanatorium four days later, on March 22, 1941.

The year 1941 also witnessed the founding of the research institute (*Forschungsinstitut*) under the direction of Caspar Huegel, which included the new Union of Museums, Archives, and Scientific Libraries (Verband der Museeun, Archive und wissenschaftliche Bücherein) led by Rudolf Spek, while academics and lay researchers were organized as themed "work communities" (*Arbeitsgemeinschaften*) that formed the German Research Community in Romania (Deutsche Forschungsgemeinschaft in Rumänien) directed by Fritz Roth.[28] The institute itself comprised seven departments: Germanist Studies (Walter Scheiner), Cultural Anthropology (*Volkskunde*) (Misch Orend), Prehistory (Fritz Roth), History (Erhard Antoni), Research into Law and Legal History (Wilhelm Klein), Racial Studies (Eckhard Hügel), and Natural Sciences (Viktor Weindel).

While Eckhard Hügel is one of the Saxon interwar eugenicists to gain in prominence after 1940, it was in particular Viktor Weindel, who also led the Transylvanian Saxon Society for Natural Sciences (Siebenbürgischer Verein für Naturwissenschaften)[29] into the *Forschungsinstitut*, who began a transformative reform of Saxon medicine, sweeping away existent associations in favor of a centralized, totalitarian approach to conceptualizing and delivering healthcare. The resulting medical approach to individual and public hygiene wholly mimicked its Third Reich counterparts, being managed by a central office of National Health, one that could overcome the

27 ZAEK, LK 103.100 (1941), 318.

28 FH, "Die deutsche Volksforschung in Siebenbürgen."

29 Weindel, "Zum Geleit!" and "Die Eingliederung des in das Forschungsinstitut der Deutschen Volksgruppe in Rumänien."

much derided individualism and fractured medical landscape of "the liberal era that did not understand health leadership [*Gesundheitsführung*] as a political duty."[30] As Wilhelm Hager (as director of the Main Office for National Health) and Viktor Weindel (as first chair of the Union of German Doctors in Romania) put it in November 1940, Schmidt's creation of the NSDAP of the German Volksgruppe and the subsequent deal with Marshal Antonescu (November 9 and 21, respectively) first laid the foundation upon which a comprehensive, pan-German system for all of Romania's Germans could be built with enough executive clout to safeguard that national community's "biological existence, its racial purity and healthy development."[31]

This new health office took the form of the highly centralized Main Office for National Health (Hauptamt für Volksgesundheit), which itself contained two suboffices, one for health services (Amt für Gesundheitswesen) the other for physical education (Amt für Leibesübungen).[32] By 1942, it came to subsume six departments, covering organization and inspection; medical provision (Gesundheitsdienst); medical insurance and hospitals; hygiene; press and propaganda; as well as racial hygiene and hereditary biology (Rassenhygiene und Erbbiologie), a department tasked with working on "measures in racial and hereditary health," and itself subsuming departments on hereditary survey, hereditary index, hereditary archive, and racial-legislative measures (Erbbestandsaufnahme, Erbkartei, Erbarchiv, and Rassengesetzliche Massnahmen).[33]

That said, it is somewhat ironic that the 1940 reorganization, as partly motivated by the increasingly autonomous Saxon position vis-à-vis the Romanian state, unwittingly also reflected proposals for a much more active medical profession first suggested in a 1919 "Report of the Saxon Medical Association on the Future Configuration of Saxon Healthcare" submitted by Siegmund.[34] This three-page-long report, having defined Saxon health-

30 Mathias, "Der Aufbau des Amtes für Volksgesundheit," 5. Also see Frank, "Das Gesundheitswesen unserer Volksgruppe."

31 Hager and Weindel, "Deutsche Ärzte!" Also see Wokalek's short overview of Saxon medical history since 1919: Wokalek, "Die Volksgesundheitliche Lage und Betreuung der Volksgruppe." This is, of course, not entirely true considering Siegmund's Welfare Committee, albeit church-based, or that Schunn's resort as Commissioner of the Nation's Organic Constitution included a Department of Health (Arbeitstelle für Volksgesundheit) under Dr. Wilhelm Hager, about which we know little.

32 Schmidt, "Der vorläufige Arbeitsrahmen."

33 Mathias, "Der Aufbau des Amtes für Volksgesundheit," 6–7.

34 See Sächsischer Ärzteverein, ed., *Gutachten des sächsischen Ärztevereins.*

care's two core competences as encompassing the quantitative "increasing of offspring" and qualitative "maintenance and improvement [*Ertüchtigung*] of offspring," set out several key methodologies, including racial hygiene alongside individual and public health.

Yet despite this ambitious agenda, the Saxon medical association's original 1902 statutes remained largely unchanged throughout the interwar period (save minor corrections in 1924 and again in 1933),[35] lumbered by the political deadlock after 1935, as well as by an ongoing conflict with the Banat Schwab Semmelweis Medical Group (Semmelweis Gruppe der Banater Ärzte) that it contained. In 1939, a new medical association was founded with Viktor Weindel as its chair, the short-lived Union of German Doctors in Romania (Bund deutscher Ärzte in Rumänien) that was superseded by the German Doctors in Romania (Deutsche Ärzteschaft in Rumänien), the creation of which was announced at the union's annual meeting in Temeschburg, in May 1941. Introducing Wokalek, Weindel justified the revamp by arguing that "the way we have governed ourselves until know, the characteristics of which were choice and elections, these days—when a tough authoritarian leadership has proven itself necessary in all other areas—has become outdated, not fit for purpose or our times."[36] Consequently, the union's executive committee resigned their posts to be replaced by Schmidt's appointments, making Wokalek chair to the new medical association, while Weindel became the deputy.

Wokalek's opening address was saturated with the usual political and ideological noises, lauding the authoritarian model and setting out the Main Office for National Health's main duties as streamlining all medical work, "supporting the hereditarily industrious [*Erbtüchtigen*], premarital counseling, prenatal and antenatal care, the battle against infant mortality, promoting the willingness to have more children, the elimination of the hereditarily diseased [*Ausmerzung der Erbkrankheiten*], preventing the inflow of foreign blood [*Verhinderung des Einströmens artfremden Blutes*], battling national plagues, tuberculosis, rickets, goiter, and the defense against other social plagues, increasing the overall productivity."[37]

35 See Siebenbürgisch-deutscher Ärzteverein, ed., "Die rechtlichen."
36 "Vollversammlung des Bundes der Deutschen Ärzte," 85.
37 Ibid. For the full text of Wokalek's speech, see Wokalek, "Der Arzt als Treuhänder der Volksgesundheit."

A year later, in opening the first issue of the Arzt im Osten medical association's new organ, Wokalek spells out the profession's duties as "keeping German blood pure, caring for the German child, and the maintenance of national health."[38] So while eugenics and racial hygiene were certainly frequently invoked in the post-1940 Schmidt era, as was the pan-German ideal of unifying all of Romanian's Germans under one coherent—and increasingly autonomous—system through an authoritarian party political framework, one that built upon but was unequivocally independent (not to say forgetful) of the interwar eugenic discourse, we know remarkably little about what specific policies were pursued or implemented unto the catastrophic collapse of August 1944.

That said, Wokalek's career is illustrative of the new, post-1940 generation of Saxon fascists and eugenicists, albeit there are many as yet unfilled gaps in his biography. Born in 1914, Wokalek studied medicine in Prague, Greifswald, Erlangen, Graz, and Berlin, where he submitted his thesis on the heredity of eye disorders in 1939.[39] Wokalek subsequently volunteered for service in the SS Leibstandarte Adolf Hitler from September to November 1939, which he formally joined in August 1940, before assuming key functions in the medical oversight of resettlement projects in Ukraine, Bessarabia, and Dobrudja in 1940 and 1941. Wokalek rapidly climbed through the ranks, with numerous deployments between 1941 and 1944—both to the front as well as the SS-Führungsamt and SS-Sanitätsamt, amongst others—and was promoted to the rank of captain (*Hauptsturmführer*) in November 1943. After being wounded in late 1943, Wokalek was sent back to Transylvania in March 1944, but we know little about what happened to him after his return.

Punctuated by these various deployments, Wokalek built and headed the new German Doctors in Romania (Deutsche Ärzteschaft in Rumänien) medical association, as well as the Main Office for National Health (Hauptamt für Volksgesundheit) from 1941 onwards. Alongside his biography, both of these organizations are very much in need of further research to establish basic institutional histories and to evaluate their overall impact on Saxon medical provisions between 1940 and 1944.

38 Wokalek, "Zum Geleit!"

39 Wokalek, *Statistische Untersuchungen.*

In his 1942 article "National Socialist Healthcare," Wokalek further elaborates on the overt National Socialist reinterpretation and management of the individual's duty to health, and the eugenic yardstick to measure it by:

> Tending to bodily health and the purity of blood are fundamental demands of the National Socialist *Weltanschauung*. Their protection is the highest duty of any National Socialist national leadership [*Volksführung*], which in order to live up to this task has created the Department for National Health [Amt für Volksgesundheit].
> [. . .] Our work is still—taking the longer view—the prerequisite for creating these formations with racially highly valuable, and physically truly healthy, people—as a purposeful national health policy [*Volksgesundheitsführung*] must anyway guarantee the eternal existence of the nation in its racial, bodily, and spiritual purity.
> [. . .] We proceed from the fundamental principle that the individual's health is no longer his private matter, but that the right of a person to his own body is surpassed by the nation's right over it. For the individual this brings not rights, but responsibilities. First of all the duty to lead his life in a way he can justify to his own health, to his family, and his nation.[40]

This text is remarkable not simply for its ravenous embrace of the totalitarian idea, but because in doing so it also concluded the debate about "reconciling" individual and racial hygiene that Siegmund first began with his article in 1901.

40 Wokalek, "Nationalsozialistische Volksgesundheitsführung." For a longer translated extract from this text, see Georgescu, "Transylvanian Saxon Eugenics."

CONCLUSIONS

This study sought to unpack how and why a eugenic discourse that had emerged under Heinrich Siegmund's tutelage in the early twentieth century was radicalized, politicized, and institutionalized through Alfred Csallner and the Self-Help during the interwar period. Csallner was characteristic of his eugenicist peers in terms of outlook and background, as he was not a medical professional like Siegmund but, like Bredt and Schunn, a priest and teacher. Similarly, Csallner perfectly illustrates the eugenic discourse's reorientation from a church-centric to a party-political approach to translating visions of racial regeneration into practice. The Self-Help's rapid rise to power revolutionized the Saxon political discourse. And while the church necessarily remained central to the eugenic discourse due to its social significance and infrastructural network, Saxon eugenics embraced Saxon fascism as the natural route to implementing a program of national renewal.

This study, therefore, also ventured to substantiate the thesis that the Self-Help must necessarily be understood as an indigenous fascist movement, even if the National Socialist franchise was adopted piecemeal towards legitimizing its bid for power in later years. Given this framework, this study also sought to investigate the Self-Help's increasingly vociferous embrace of eugenics and racial hygiene as quintessential tools in the quest to reinvent, re-enchant, and regenerate the Saxon nation. Equally important, the Self-Help's eugenic discourse, as it was defined and refined by Wilhelm Schunn and Alfred Csallner, was not limited to rhetoric alone. Indeed, the fascist reinvention of the national neighborhoods—even if only patchily and almost exclusively in urban environments unto 1940—constituted a remarkably ambitious bioengineering project.

In short, the research offered here strove to persuasively argue that an interwar ethnic minority—in this case the Transylvanian Saxons—could define, deliberate, legislate, and even execute a eugenic agenda. It also strove, therefore, to illustrate why an ethnic minority would embrace a eugenic worldview seen to offer a novel ideological and practical tool with which to re-homogenize and re-entrench a redefined sense of self behind the race-hygienic walls of a eugenic fortress. This realization unavoidably raises the specter that many more such eugenic movements might have emerged amongst other ethnic minorities. Their excavation and comparison to the Saxon experiment with eugenics will undoubtedly make an invaluable contribution to our understanding of interwar nation and identity formation.

That said, the analysis of Saxon eugenics advanced here wants to be understood as both a self-contained (if certainly not exhausted) case study, as well as a springboard; to have not only shed light on a discourse fundamental to, if neglected by Saxon historiography, but to have also illustrated the viability of a wide array of further studies, of future research—as rewarding as it is necessary. If nothing else, though, virtually any investigation of socioeconomic, demographic, and even migratory trends in interwar Transylvania would (and, indeed, should) benefit from a quantitative analysis of the vast statistical databases amassed by Csallner and his LSS. But the in-depth study of Saxon eugenics has far more to offer than empirical data. Given its substantial influence on Saxon self-perception and politics, it can inform a wide variety of existing research paradigms that must necessarily incorporate the eugenic discourse if they are to be authoritative.

The comprehensive excavation of Saxon eugenics as it emerged and evolved over three distinctive periods between 1885 and 1944 hence raises a host of new research questions, and offers new perspectives on established ones. Of the numerous lines of enquiry that stand to benefit from further research, the following merit particular attention: for one, the characterization of Fritz Fabritius's Self-Help as a distinctly Saxon permutation of "generic fascism," and the complementary question of whether other ethnic minorities hosted comparable fascist discourses (or even adopted the Saxon Self-Help model). Similarly important, the illumination of Alfred Bonfert's Self-Help Work Teams (SAM) and the Volunteer Work Camps both in terms of their stated aim to engineer a "new man," and in comparison to those institutionalized by other fascist discourses, such as the Romanian legionary movement. Such a research project would provide much needed insights into the socioeconomic makeup of the Self-Help's youth movement and its activities, in addition to addressing the degree of ideological connectivity, and level of interaction, between the SAM and its Romanian counterpart. A third research project that would help to further define and locate Saxon eugenics within the wider international context that informed it would focus on the intimate ties that the Saxons nurtured with German and Austrian eugenicists, compared, for example, to the absence of any substantive interaction with the Hungarian and Romanian eugenic movements. But any investigation of these international links would have to discuss an enticing triad of issues, namely: the transfer of ideas through Saxon students returning to Transylvania; the process by which these ideological imports were appropriated and adapted to Saxon needs; and the nature of ideological, fiscal, and institutional support that Saxon eugenicists received, in particular, from Germany (before and after 1933). While these issues have been discussed here, a considerably more elaborate study would make a substantial contribution to our understanding of not only Saxon eugenics, but also accentuate that of such eminent and widely discussed figures such as Alfred Ploetz, Eugen Fischer, and Viktor Lebzelter.

Saxon eugenics constituted but one of the currents subsumed by the ideological sea change that swept across Europe with the rise of eugenics, which is particularly evident in the life and work of Heinrich Siegmund. The founding father of Saxon eugenics, Siegmund's virtually complete absence from Saxon historiography is beyond perplexing, particularly so due to the ready availability of a vast archival base in the Romanian National Archives (holding his personal papers), the Teutsch Haus (housing the church archive and the materials left by his Welfare Committee), in addition to his hundreds of publications. As this study has repeatedly argued, his subjection to a detailed evaluation will necessarily revolutionize current discourses on the history of Saxon medicine and the National Protestant church alike.

This can surely be but the beginning of historiography's interest in a eugenic discourse as varied and faceted as that produced by the Transylvanian Saxons, one that accommodated three distinct phases between 1885 and 1944, each feeding into and on a remarkable range of local, regional, and international networks perpetually reconfigured to suit an ever-changing geopolitical reality. In that sense, Schmidt's wartime *Volksgruppe* marked the conclusion of a discourse begun by Siegmund forty years previously, at the hands of a new, Berlin-orchestrated eugenic program, a generational change introducing a largely new set of protagonists. Fundamentally, though, it also signaled the beginning of the end for the interwar architects of the eugenic fortress.

BIBLIOGRAPHY

Archival Material

Central Archive of the National Protestant Church A.B. in Romania (Zentralarchiv der Evangelischen Kirche A.B. in Rumänien), Sibiu:

Landeskonsistorium Archive (Actela Cancelar): Inv. 1923–1942;

Personal Papers (Fonduri şi lăsăminte personale): Inv. 503: Wilhelm Bruckner: 538–39, 541, 548–49, 552–53, 562, 566, 569, 573, 575–76, 585, 721, 725, 795–96, 798, 848, 850–51, 866, 869, 879, 881–82, 886, 897–98, 903, 951, 1068, 1186, 1188, 1217, 1243–44, and 1247; Inv. 526: Wilhelm Schunn: 1–18; Inv. 539: Julius Ernst Gyurgyewich.

Eötvös Loránd University, Budapest, Philosophy Faculty Yearbooks, Alfred Csallner's student papers, 1913–14 and 1917–18.

German Federal Archive (Bundesarchiv), Berlin:

NS 2: Race and Settlement Office (Rasse- und Siedlungshauptamt): 137, 161–62, 217, 227, and 294;

R 49: Reich Commissar for the Strengthening of Germanism (Reichskommissar für die Festigung deutschen Volkstums): 12, 25, 71, 3040, 3048, 3059, 3063, 3066, 3069, 3073, 3083, and 3127;

R 59: Ethnic German Liason Office (Volksdeutsche Mittelstelle): 323–25, 328, 331–32, 336, 343, 347, 355–56, 369, 375, 376–78, 385–86, and 393;

R 1509: Reich Kinship Office (Reichssippenamt): 1552, 1622, 1624, 1763, 1774–76, and 1837;

R 4909: German Foreign Institute (Deutsches Auslandswissenschaftliches Institut): 2761, 2762, 4343, 9251, 9254, 10959, and 11481.

Romanian National Archives (Arhivele Naţionale ale României), Sibiu:

Brukenthal Collection (Colecţia Brukenthal): Inv. 106: L.1–8.15; Inv. 107: LL.1–29.38–107 (Heinrich Siegmund's personal papers); Inv. 125: Z 1–7.196; and Z 1–7.200.1944;

Saxon National Council (Consiliul Naţional Săsesc): Inv. 1919: 29; Inv. 1933: 17–20, 22, and 36–7; Inv. 1934: 4–5; Inv. 1935: 4, 8, and 33; Inv. 1936: 6–7, 39;

Association of Germans in Romania (Comunitatea Germanilor din România):

Inv. A: Central Office (Oficiul Central): I.1–4;

Inv. D: Local Office (Oficiul Provincial):

XIII: National Neighborhoods (Oficiul pentru Vecinătăţi):

Central Office (Oficiul Central): 205–336;

Sibiu Neighborhoods (Vecinate Sibiu): 337–681;

XIV: National Office for Statistics, Demographic Policy, and Genealogy (Oficiul pentru Statistică, Politică Demografică şi Genealogie): 682–851;

Inv. K: Collections (Colecţionate): 13, 16–19, and 23;

Supplementary Inventory (Inventar Suplimentar): 19.

Siebenbürgen Institute, Gundelsheim, Personalities (Persöhnlichkeiten): Folder B: Johann Bredt; Folder C: Alfred Csallner; Folder S: Heinrich Siegmund.

University of Vienna, Nationalien der Philosophischen Fakultät der Universität Wien, Alfred Csallner's student papers, 1917.

Transylvanian Saxon Periodicals

Der Arzt im Osten 16 (1942)–18 (1944).

Der Aufbau: Begeisterung ist nichts ohne die Tat 1 (1923)–18 (1942).

Der Aufbau: Monatsschrift für Volksglauben, Volkskultur, Volkswirtschaft 1 (1934)–2 (1935).
Der Aufbau: Wochenschrift für die Deutsche Volksgemeinschaft in Rumänien 1 (1935)–.
Blätter für Kinderfrohe 2 (1929)–3 (1930).
Deutsche Forschung im Südosten: Forschungsinstitut der Deutschen Volksgruppe in Rumänien 1 (1942)–3 (1944).
Deutsche Politische Hefte 6 (1926)–7 (1927).
Deutsche Politische Hefte aus Grossrumänien 1 (1921)–6 (1926).
Evangelischer Fürsorger: Beilage der Kirchlichen Blätter aus der evangelischen Landeskirche A.B. in Siebenbürgen 1 (1924)–16 (1940).
Fürsorge-Blätter der evangelischen Landeskirche A.B. in Siebenbürgen 1–19, n.d.
Die Karpathen: Halbmonatsschrift für Kultur und Leben 1 (1907)–7 (1913).
Kirchliche Blätter aus der evangelischen Landeskirche A.B. in Siebenbürgen: Evangelische Wochenschrift für die Glaubensgenossen aller Stände 17 (1925), 24 (1932), 26 (1934)–28 (1936), 30 (1938).
Klingsor: Siebenbürgische Zeitschrift 1 (1924)–16 (1939).
Korrespondenzblatt des Vereines für siebenbürgische Landeskunde 1 (1878)–40 (1917), 42–43 (1919–1920), 45 (1922), 47 (1924), 49 (1926)–52 (1929).
Medizinische Zeitschrift: Fachblatt der deutschen Ärzte in Rumänien 1 (1927)–15 (1941).
Mitteilungen der Arbeitsgemeinschaft für Naturwissenschaften 91–92 (1941–42).
Ostdeutscher Beobachter: Kampfblatt für das ehrlich arbeitende Volk 11 (1932)–13 (1934).
Ostland: vom geistigen Leben der Auslandsdeutschen 3 (1928)–5 (1930).
Sachs' halte Wacht: Zeitschrift heimattreuer Siebenbürger Sachsen und ihrer Freunde 1 (1927)–2 (1928).
Schule und Leben: Fachzeitschrift des Siebenbürgisch-Sächsischen Lehrerbundes, 1936–1938.
Selbsthilfe 1 (1922)–11 (1932).
Selbsthilfe für Kinderfrohe 3, no. 4 (1930).
Siebenbürgische Vierteljahresschrift: Korrespondenzblatt des Vereins für Siebenbürgische Landeskunde 54 (1931)–64 (1941).
Verhandlungen und Mitteilungen des Siebenbürgischen Vereins für Naturwissenschaften zu Hermannstadt 68–69 (1918–19)–89–90 (1939–40).
Volk im Osten: politische Monatsschrift für das Deutschtum im Osten 1 (1934)–4 (1937).
Volk im Osten: Zeitschrift des Südostens 1 (1940)–1 (1940).
Volk und Heimat: Deutsche Monatschrift für Rumänien 1 (1937)–3 (1939).
Volk und Rasse—Volk und Raum: Schriften des Amtes für Erbbiologie beim Kulturamt der Deutschen in Rumänien 1 (1934)–11 (1940).
Volksgesundheit: Gemeinschaftliche Monatsschrift für deutsch-ungarische Kulturpolitik 1 (1902)–9 (1911).

German Periodicals

Archiv für Rassen- und Gesellschafts-Biologie: Einschliesslich Rassen- u. Gesellschaftshygiene 4 (1907)–37 (1943).
Auslandsdeutsche Volksforschung 1 (1937)–2 (1938).
Deutsche Erde: Beiträge zur Kenntnis deutschen Volkstums allerorten und allerzeiten 1 (1902)–13 (1914–15).
Deutsches Archiv für Landes- und Volksforschung 1 (1937)–5 (1941), 7 (1943)–8 (1944).
Deutschtum im Ausland: Zeitschrift des Deutschen Ausland-Instituts Stuttgart 25 (1942)–27 (1944).
Jahrbuch für Auslandsdeutsche Sippenkunde 1 (1936)–2 (1937).
Nationalsozialistische Monatshefte 1 (1930)–12 (1941).
Südostdeutsche Forschungen 1 (1936)–8 (1943).
Volksforschung 3 (1939–40)–6 (1942–43).
Zeitschrift für Rassenphysiologie: Mitteilungen der Deutschen Gesellschaft für Blutgruppenforschung 1 (1929)–13 (1943).

Published Sources

"Amtlicher Bericht über die Volksratssitzung vom 21 April 1934." *OB*, April 28, 1934, 3.
"An die deutsch-sächsische Bevölkerung von Hermannstadt und weiterhin im ganzen romänischen Vaterlande!" *Selbsthilfe* 3 (1922): 1.
"Aufbruch." *OB*, August 27, 1932, 1–2.
"Die Auflösung der NEDR im Amtsblatte verlautbart." *SDT*, July 13, 1934, 1.
"Aus den Verhandlungen der Bukarester Pfarrkonferenz." *Kirchliche Blätter* 7 (1928): 64–66.
"Ausblick nach dem Sachsentag." *OB*, October 14, 1933, 1.
"Ein Ausschnitt aus den bisherigen Wahlerfolgen." *OB*, November 11, 1933, 3.
"Ausserordentliche Generalversammlung." *Selbsthilfe* 16 (1929): 1–2.
"Bisher geleistete Arbeit." *Selbsthilfe* 5 (1925): 2.
"Bürger-Abendrot." *OB*, October 28, 1933, 4.
"Durch den Sachsentag zum Neuaufbau unserer Volksorganisation." *OB*, May 20, 1933, 1–2.
"Ein Ehrentag des Deutschtums in Bessarabien." *OB*, March 31, 1934, 4.
"Exekutive?!" *OB*, April 21, 1934, 1.
"Frieden mit der Kirche." *OB*, March 24, 1934, 3.
"Geschäftsordnung der Firma 'Selbsthilfe,' Unternehmung zur Vermittlung und Finanzierung von Immobilien Fr. Fabritius & Co, Kommanditgesellschaft in Hermannstadt." *Selbsthilfe* 12 (1928): 4–6.
"Die grosse Volksversammlung der NEDR in Mediasch am 18. März 1934: Die grösste Kundgebung seit dem Sachsentag!" *OB*, March 31, 1934, 5–6.
"Die 'Kronstädter Zeitung' [vom 9 September 1930]." *Selbsthilfe*, October 15, 1930, 3.
"Kulturpreis für Heinrich Bredt und Alfred Csallner: Dem Altersforscher und dem Bevölkerungswissenschaftler." *Siebenbürgische Zeitung*, June 15, 1982, 4.
"Die Kundgebung der NEDR im Burzenland: Das Bekentniss zu Heimat, Volk und Vaterland." *OB*, June 30, 1934, 2–4.
"Kundgebung der NEDR in Kronstadt." *OB*, June 30, 1934, 1.
"Liste der ausbezahlten Beträge bis Ende August 1928." *Selbsthilfe* 12 (1928): 6.
"Mitgliederversammlung der O.G. Hermannstadt." *Selbsthilfe*, November 1, 1930, 4–5.
"Nachricht." *Selbsthilfe* 9 (1927): 4.
"Der Nationalsozialismus (Grundsätze und Bewegung)." *Selbsthilfe*, October 28, 1931: 1-2.
"Die NEDR baut auf, ihre Gegner reissen nieder: Die Massenversammlung der NEDR am 15 April 1934 in Hermannstadt." *OB*, April 21, 1934, 4–6.
"Der neue Sozialismus! Teil I." *Selbsthilfe* 4 (1922): 3–4.
"Organisationsrichtlinien der Nationalsozialistischen Selbsthilfebewegung der Deutschen in Rumänien (NSDR)." *OB*, August 27, 1932, 4.
"Pressestimmen über die 'Selbsthilfe.'" *Selbsthilfe*, October 15, 1930, 2–3.
"Richtlinien der 'Selbsthilfe,' Unternehmung zur Vermittlung und Finanzierung von Immobilien Fr. Fabritius & Co, Kommanditgesellschaft in Hermannstadt." *Selbsthilfe* 12 (1928): 3–4.
"Der Sachsentag im Zeichen des Sieges der NSDR: Das Volk hat gesprochen, Brandsch musste schweigen." *OB*, October 7, 1933, 2.
"Die sächsische Pfarrerfrage." *Selbsthilfe*, May 14, 1932, 1.
"Das sächsische Volksprogramm: Ein Entwurf der Selbsthilfe für den im Herbst 1932 abzuhaltenden Sachsentag." *OB*, August 27, 1932, 1–3.
"Selbsthilfe-Arbeitsmannschaft." *OB*, September 10, 1932, 3.
"Spiel mit dem Volk." *OB*, April 28, 1934, 1.
"Unser Vormarsch: Die Gautagung in Hermannstadt." *Selbsthilfe*, December 15, 1930, 2–4.
"Veranstaltungen des Gaues Hermannstadt der Selbsthilfe: 15 und 16 November 1930." *Selbsthilfe*, November 15, 1930, 5.
"Vereinbarung zwischen ev. Landeskirche AB in Rumänien und NEDR." *OB*, March 24, 1934, 3–4.
"Die Verköterung der Menschheit." *Selbsthilfe* 11 (1928): 4.
"Verschiedenes." *Selbsthilfe* 9 (1927): 4.

"Volksgenossen." *Selbsthilfe* 3 (1922): 3–4.
"Vollversammlung des Bundes der Deutschen Ärzte in Rumaenien am 10. Mai 1941 in Temeschburg." *Medizinische Zeitschrift* 4–5 (1941): 85–87.
"Vor der Entscheidung." *OB*, February 4, 1933, 1–2.
"Was wollen wir auf dem Sachsentag." *OB*, May 13, 1933, 1.
"Woran glauben Sie? 16 Fragen an bedeutende und bekannte Persönlichkeiten, Heute: Pfarrer Professor Alfred Csallner." *Deutsche Nationalzeitung* 46 (1985): 7.
"Zum Abschluss." *Selbsthilfe*, January 1932, 1.
"Zur Klarstellung der Lage: Ein Wort an alle deutschen Volksgenossen." *OB*, July 28, 1934, 3.
A.K. "Unsere Politik." *Selbsthilfe*, April 30, 1932: 1.
Amlacher, Albert. *Rumes: Aus Vergangenheit und Gegenwart einer siebenbürgisch-sächsischen Dorfsgemeinde*. Hermannstadt: n.p., 1912.
Antohi, Sorin. "From Metaphysics to Biology: The Dark Side of Ethnic Ontology in Greater Romania." Paper presented at the conference "Greater Romania's National Projects: Ideological Dilemmas, Ethnic Classification, and Political Instrumentalization of Ethnic Identities," Oxford Brookes University, April 10–12, 2008. http://www.pulse-project.org/node/78.
Baier, Hanelore. "Politische Initiativen und Organisationen der siebenbürger sachsen in der Zeitspanne 23. August 1944–Februar 1949." *Forschungen zur Volks- und Landeskunde* 1–2 (1996): 55–80.
Balling, Mads Ole. Review of *Politische Strukturen und Strömungen bei den Siebenbürger Sachsen 1919–1933*, by Harald Roth. *ZfSL* 2 (1995): 212–13.
Bashford, Alison, and Philippa Levine, eds. *The Oxford Handbook of the History of Eugenics*. Oxford: Oxford University Press, 2012.
Beer, Mathias, and Gerhard Seewann, eds. *Südostforschung im Schatten des Dritten Reiches: Institutionen—Inhalte—Personen*. München: Oldenburg Wissenschaftsverlag, 2004.
Behrwerth, Fritz. "Dr. Heinrich Siegmund." *Südostdeutsche Vierteljahresblätter* 2 (1970): 104–8.
Bell, Richard. "An ihren Früchten sollt ihr sie erkennen." *Evangelischer Fürsorger* 8 (1937): 30–31.
Bergleiter, Misch. "Deutsche Weihnacht." *Selbsthilfe* 25 (1929): 1–2.
———. "Vorspruch [Gedicht] zur Führertagung der 'Selbsthilfe' am 4 Ostermonds 1930, gehalten von Misch Bergleiter." *Selbsthilfe* 5 (1930): 1–2.
Binder, Ludwig. "Die Evangelische Landeskirche A.B. in Rumänien, 1920–1944." In *Siebenbürgen zwischen den beiden Weltkriegen*, edited by Walter König, 237–64. Köln: Böhlau, 1994.
Blackbourn, David. *The Conquest of Nature: Water, Landscape, and the Making of Modern Germany*. London: Jonathan Cape, 2006.
Blamires, Cyprian, and Paul Jackson. *World Fascism: A Historical Encyclopedia*, vols. I–II. Santa Barbara: ABC-CLIO, 2006.
Blomqvist, Anders, Constantin Iordachi, and Balázs Trencsényi, eds. *Hungary and Romania Beyond National Narratives: Comparisons and Entanglements*. Frankfurt am Main: Peter Lang, 2013.
Böhm, Johann. "Anmerkungen zu einem Plagiatsverdacht." *HJS* 1 (2008): 61–64.
———. *Die Deutschen in Rumänien und das Dritte Reich, 1933–1940*. Frankfurt am Main: Peter Lang, 1999.
———. *Die Deutschen in Rumänien und die Weimarer Republik, 1919–1933*. Ippesheim: AGK Verlag, 1993.
———. *Die Gleichschaltung der deutschen Volksgruppe in Rumänien und das "Dritte Reich," 1941–1944*. Frankfurt am Main: Peter Lang, 2003.
———. *Hitlers Vasallen der Deutschen Volksgruppe in Rumänien vor und nach 1945*. **Frankfurt: Peter Lang, 2006.**
———. *Das Nationalsozialistische Deutschland und die deutsche Volksgruppe in Rumänien, 1936–1944: Das Verhältnis der deutschen Volksgruppe zum Dritten Reich und zum rumänischen Staat sowie der interne Widerstreit zwischen den politischen Gruppen*. Frankfurt am Main: Peter Lang, 1985.
———. *Nationalsozialistische Indoktrination der Deutschen in Rumaenien 1932–44*. Frankfurt am Main: Peter Lang, 2008.
———. "Die politische Entwicklung der deutschen Minderheit in Rumänien von 1918 bis 1944." *Revista Istorică* 5–6 (2000): 345–69.

———. "Von der völkischen Verblendung zum Nationalsozialismus, Teil II." *HJS* 2 (2006): 55–69.

Böhm, Johann, and Dieter Braeg, eds. *Dr. Viktor Glondys: Tagebuch – Aufzeichnungen von 1933 bis 1949*. Dinklage: AGK Verlag, 1997.

Böhm, Johann, and Klaus Popa. *Vom NS-Volkstum- Zum Vertriebenenfunktionär: Die Gründungsmitglieder Des Südostdeutschen Kulturwerks München Und Der Landsmannschaften Der Deutschen aus Rumänien, Ungarn und Jugoslawien*. Frankfurt am Main: Peter Lang, 2014.

Bonfert, Alfred. "Das Arbeitslager in Meschendorf: 17–23 August 1931: Teil I." *Selbsthilfe*, September 26, 1931, 6.

———. "Das Arbeitslager in Meschendorf: 17–23 August 1931: Teil II." *Selbsthilfe*, October 28, 1931, 1–2.

Bredt, Heinrich. "Volkskörperforschung in Siebenbürgen." *Jahrbuch für Auslandsdeutsche Sippenkunde* 1936, 141–142.

Bredt, Johann. "Der Schwindsucht in unserer Gemeinde: Entwurf eines einleitenden Vortrages für die Fürsorgewoche." *Evangelischer Fürsorger* 12 (1929): 132–39.

———. *Vererbungskunde und Katechismusbehandlung*. Hermannstadt: Honterus, 1932. Also published as: "Vererbungskunde und Katechismusbehandlung: Teil I." *Evangelischer Fürsorger* 1 (1932); and "Vererbungskunde und Katechismusbehandlung: Teil II." *Evangelischer Fürsorger* 2 (1932).

———. *Volkskörperforschung*. Breslau: Ferdinand Hirt, 1930.

———. *Windau in Gegenwart und Vergangenheit: Eine Ortsgeschichte auf Grundlage der Volkskörperforschung*. Bistriz: Buchdruckerei Gustav Zikeli, 1929.

Bredt, Johann, and Günter Litschel. *Windauer Heimatbuch*. Schweinfurt: Schweinfurter Druckerei und Verlagsanstalt, 1971.

Bucur, Maria. *Eugenics and Modernisation in Interwar Romania*. Pittsburgh: Pittsburgh University Press, 2002.

Burleigh, Michael. *Germany Turns Eastwards: A Study of* Ostforschung *in the Third Reich*. London: Pan Books, 2002.

Burleigh, Michael, and Wolfgang Wippermann. *The Racial State: Germany 1933–1945*. Cambridge: Cambridge University Press, 1991.

Capesius, Bernhard. "Die grösste Gefahr?" *Kronstädter Zeitung*, September 20, 1935.

Ciobanu, Vasile. *Contribuţii la cunoaşterea istoriei saşilor Transilvanei 1918–1944* [Contributions to the knowledge of Transylvania's Saxons, 1918–1944]. Sibiu: Editura Hora, 2001.

———. *Identitatea culturală a germanilor din România în perioada interbelică* [The cultural identity of Germans in Romania in the interwar period]. Bucureşti: Editura Muzeului Naţional al Literaturii Române, 2013.

———. "Neue Quellen zur Geschichte der Siebenbürger Sachsen in den Jahren 1918–1944." *Forschungen zur Volks- und Landeskunde* 2 (1980): 49–58.

———. "**Relaţiile culturale dintre grupurile de germani din România în primul deceniu interbelic**" [Cultural relations between the German groups in Romania in the first interwar decade]. *Ţara Bârsei* 10 (2011): 155–61.

Cleminson, Richard. *Catholicism, Race, and Empire: Eugenics in Portugal, 1900–1950*. Budapest: Central European University Press, 2014.

Collmer, Paul. *Fürsorge als völkische Selbstbehauptung: Dargestellt am Beispiel des Fürsorgewesens der Siebenbürger Sachsen*. Berlin: Carl Heymanns Verlag, 1936.

Costa Pinto, António, ed. *Rethinking the Nature of Fascism: Comparative Perspectives*. London: Palgrave Macmillan, 2011.

Csallner, Alfred. "Achthundert Jahre . . . !" *Selbsthilfe*, August 8, 1931, 4–5.

———. *Ahnenpass*. Hermannstadt: Bruckner, 1940.

———. *Anleitung zur Bearbeitung der Fragebögen des deutschen Volkes in Rumänien*. Manuscript, 1939. NA Sibiu, CGR, D.XIV.835, pp. 1–16.

———. *Die Arbeitsstellen für Statistik, Bevölkerungspolitik und Sippenwesen: Ein Wegweiser*. Hermannstadt: Volksgemeinschaft der Deutschen in Rumänien, 1935.

———. "Das Auslandsdeutschtum und Rassen- und Erbgesundheitspflege im Reich." *Volk und Heimat* 1 (1938): 11–14.

———. *Der Baruch und andere Erzählungen aus Siebenbürgen.* Sankt Michael: Bläschke, 1980.
———. "Beispiele für die wachsende Kinderbeschränkung unter uns Sachsen." *Blätter für Kinderfrohe* 1–2 (1929): 7.
———. *Bestandsaufnahme: Auswertung der ländlichen Aufnahmebögen.* Hermannstadt: G.A. Haiser, 1940.
———. *Betrachtungen zur Lage.* N.p., 1928. Also published in *Kirchliche Blätter,* 1928.
———. "Binnenwanderung im Sachsenland." *Evangelischer Fürsorger* 12 (1927): 235–36.
———. *Der Bistritzer deutsche Akademikerstand, gesellschaftsbiologisch betrachtet.* Hermannstadt: Honterus, 1934. Also published in *Volk und Rasse—Volk und Raum* 3 (1934); *Aufbau: Monatschrift* 5 (1934).
———. "Deutsche Verluste im Nösnerland." *Bistritzer Kalender,* 1921, 48–59.
———. *Die deutschen Siedlungen in Rumänien nach der Bestandsaufnahme vom 3 November 1940.* Hermannstadt: Krafft & Drotleff, 1940.
———. "Du und Dein Volk." *Selbsthilfe,* June 15, 1931, 5–6.
———. *Die Erforschung der Abwanderung vom Dorf, dargestellt am Beispiel Grossalisch.* Hermannstadt: Honterus, 1940. Also published in *Volk und Rasse—Volk und Raum* 11 (1940); *Evangelische Fürsorger* 7–10 (1940).
———. "Erklärung." *OB,* November 11, 1933, 3.
———. "Des Esels Trost." *Aufbau: Monatschrift* 2 (1934): 14–15.
———. "Die evangelischen Zigeuner des Bistritzer Kirchenbezirks und die Statistik der Sachsen." *Bistritzer Deutsche Zeitung,* December 24, 1919, 5.
———. "Fackelträger." In *Kleine Aufsätze,* by Alfred Csallner. Hermannstadt: Selbstverlag, 1928. Also published in *Kirchliche Blätter,* 1928.
———. *Ein falscher Trost.* Hermannstadt: Honterus, 1937. Also published in *Volk und Rasse—Volk und Raum* 6 (1937); *Volk und Heimat* 2–3 (1937).
———. "Familienbuch—Erbarchiv." *Evangelischer Fürsorger* 1 (1939): 1–3.
———. "Förderung des Kinderreichtums durch unsere Kirche." *Evangelischer Fürsorger* 7 (1927): 174–78.
———. "Die Frage der Mischehe: Ein Brief." *Evangelischer Fürsorger* 1 (1940): 2–4.
———. "Geburteneinsturz und Familienstimmrecht." *Blätter für Kinderfrohe* 1–2 (1929): 1–5.
———. "Die grösste Gefahr: Ergebnisse einer erb- und gesellschaftsbiologischen Untersuchung über die Siebenbürger Sachsen." *Volk im Osten* 8 (1935): 249–57.
———. *Haben wirklich die armen Bauern weniger und die reichen mehr Kinder: Sind unsere Gemeinden wirklich überbevölkert?* Hermannstadt: Krafft & Drotleff, 1937. Also published in *SDT,* March 3–4, 1937.
———. *Irrwege.* Hermannstadt: Emil Bruckner, 1939.
———. *Der Jäger bei den Zwergen.* Sankt Michael: Bläschke, 1981.
———. "Die Kinderbeschränkung im Nösnergau." *Selbsthilfe der Kinderfrohen* 4 (1930): 13–4.
———. "Laien in die volksbiologische Forschung!" *Volk und Heimat* 8 (1938): 13–19.
———. "Das Landesamt für Statistik und Sippenwesen der Deutschen Volksgemeinschaft in Rumänien." Manuscript, October 10, 1939. NA Sibiu, CGR, Inventar Suplimentar, D.XIV.19.
———. "Lebensklammern unseres Volkes." *SDT,* November 19, 1927, 4.
———. "Luxuskinder." *Kirchliche Blätter* 48 (1927): 320–21.
———. "Das Märchen von Königen und Köhlersleuten." *Der Aufbau: Monatschrift* 7 (1934): 13–17.
———. *Der Mediascher deutsche Akademikerstand, gesellschaftsbiologisch betrachtet.* Mediasch: Feder, 1932. Also published in *Mediascher Zeitung,* January, 23–24, 1932.
———. *Meine wissenschaftlichen Arbeiten.* Manuscript, 1944. NA Sibiu, Brukenthal, Inv. 125, Z 1-7.200.1944.
———. *Meine wissenschaftlichen Arbeiten.* Traunau: Selbstverlag, 1975.
———. *"Mer welle bleiwen, wat mer senn!"—Wirklich?* Hermannstadt: Honterus, 1935. Also published in *Volk und Rasse—Volk und Raum* 4 (1935); *Aufbau: Monatschrift* 8–9 (1935).
———. "Die Mischehen in den siebenbürgisch-sächsischen Städten und Märkten: Teil I." *Auslandsdeutsche Volksforschung* 1 (1937): 225–55.

———. "Die Mischehen in den siebenbürgisch-sächsischen Städten und Märkten: Teil II." *Auslandsdeutsche Volksforschung* 2 (1938): 14–36.
———. "Die Mischehen in den siebenbürgisch-sächsischen Städten und Märkten: Teil III." *Auslandsdeutsche Volksforschung* 2 (1938): 227–60.
———. *Mutsek: Erzählungen*. Manuscript, 1976. Siebenbürgen-Institut Archiv, Ordner C, Alfred Csallner.
———. "Neujahresnacht." In *Kleine Aufsätze,* by Alfred Csallner. Hermannstadt: Selbstverlag, 1928.
———. "Offener Brief an alle, die Führer sein wollen in unserem Volke." *Selbsthilfe,* February 1, 1931: 2–4. Also published in *Bistritzer Deutsche Zeitung,* January 9, 1931; *Großkokler Bote,* January 11, 1931; and *Mediascher Zeitung,* January 24, 1931.
———. "Paarungssiebung unter den siebenbürgisch-sächsischen Bauern." *Verhandlungen und Mitteilungen des Siebenbürgischen Vereins für Naturwissenschaften zu Hermannstadt,* 1935–36, 84–106. Also published in *Volk und Rasse—Volk und Raum* 7 (1937).
———. Review of *Die Deutschen Bauern des Burzenlandes,* by Alfred Hermann. *Südostdeutsche Forschungen,* 1938, 446–48.
———. Review of *Deutschen Dämmerung in Siebenbürgen,* by Heinrich Siegmund. *OB,* August 13, 1932, 2.
———. *Rottenholz und die Rottenholzer Grosseltern*. Traunreut: Selbstverlag, 1976.
———. "Die rumäniendeutsche Bestandsaufnahmen von 1937/38 und 1940." *Südostdeutsche Vierteljahresblätter* 3 (1997): 216–18.
———. "Die sächsische Kinderbeschränkung in den Landgemeinden des Nösnerlandes." *Bistritzer Deutsche Zeitung,* September 29, 1920, 2.
———. "Der sächsische Pfarrer als Guttempler." In *Kleine Aufsätze,* by Alfred Csallner. Hermannstadt: Selbstverlag, 1928. Also published in *Aufbau: Begeisterung* 31 (1928).
———. "Sächsisches Volkswachstum." *Bistritzer Deutsche Zeitung,* January 3, 1919, 2.
———. *Schädigungen der deutschen Volksgruppe in Rumänien durch den Staat.* Manuscript, 1940. German National Library, Leipzig.
———. "Schul- und Lebensleistungen siebenbürgisch-sächsischer Bauern." *Verhandlungen und Mitteilungen des Siebenbürgischen Vereins für Naturwissenschaften zu Hermannstadt,* 1937–38, 23–34. Also published in *Volk und Rasse—Volk und Raum* 10 (1939).
———. "Die Schüler der Bestritzer evangelischen-sächsischen Schulen." *Bistritzer Deutsche Zeitung,* September 19, 1919, 4.
———. *Eine Schülerreise nach der Bukowina.* Bistritz: Theodor Botschar, 1913.
———. "'Selbsthilfe.'" *Blätter für Kinderfrohe* 1–2 (1929): 8–10.
———. *Sippschaftstafeln: Roseln- (Ruja).* Hermannstadt: Selbstverlag, 1927. Also published in *Kultur und Leben: Monatsschrift für Kulturgeschichtliche und biologische Familienkunde* 12 (1927).
———. "Spargemeinschaften Gründen!" *Evangelischer Fürsorger* 4 (1930): 13–5. Also published in *Blätter für Kinderfrohe* 1 (1930).
———. "Statistisches über den Bistritzer sächsischen Gewerbestand." In *Kleine Aufsätze,* by Alfred Csallner. Hermannstadt: Selbstverlag, 1928. Also published in *Deutsche Politische Hefte aus Großrumänien: Mitteilungen zur Kenntnis des Deutschtums in Großrumänien* 11–12 (1925); and *Bistritzer Kalender,* 1922.
———. "Steigen und Sinken der sächsischen Volkszahl im Nösnerland während der letzten anderthalb Jahrhunderte." *Bistritzer Deutsche Zeitung,* January 10, 1919, 2.
———. "Über die Wechselbeziehungen zwischen Begabung, Wohlstand, gesellschaftlicher Stellung und Kinderzahl: Ergebnisse einer Schulkinderuntersuchung." *SDT,* January 11, 1933, 5.
———. *Über volksbiologische Forschungen in siebenbürgisch-sächsischen Dörfern.* Hermannstadt: Botschner, 1939. Also published *Volk und Heimat* 2 (1939).
———. *Um der Zukunft Willen: Volksstück in drei Akten aus der Vergangenheit der Siebenbürger Sachsen.* Bistritz: Gustav Zikeli, 1927.
———. "Unsere einzigen Kinder." In *Kleine Aufsätze,* by Alfred Csallner. Hermannstadt: Selbstverlag, 1928.

———. "Unsere Verluste durch die Kinderbeschränkung." In *Kleine Aufsätze*, by Alfred Csallner. Hermannstadt: Selbstverlag, 1928. Also published in *Bistritzer Kalender*, 1922.
———. "Der völkische Charakter der Gemeinden Mönchsdorf, Moritzdorf, Tatsch und Ungersdorf." *Bistritzer Deutsche Zeitung*, September 13, 1920, 3.
———. "Die volksbiologische Forschung unter den Siebenbürger Sachsen und ihre Auswirkungen auf das Leben dieser Volksgruppe." *Deutsches Archiv für Landes- und Volksforschung* 3 (1938): 688–736.
———. *Die volksbiologische Forschung unter den Siebenbürger Sachsen und ihre Auswirkung auf das Leben diese Volksgruppe*. Leipzig: S. Hirzel, 1940.
———. "Die volksbiologische Lage der Siebenbürger Sachsen." *Medizinische Zeitschrift* 10 (1936): 345–62.
———. "Vom Mindestbedarf an Pfarrern und Lehrern und einer neuen Lösung der Besoldungs- und Bedeckungsfrage." In *Kleine Aufsätze*, by Alfred Csallner. Hermannstadt: Selbstverlag, 1928. Also published in *Kirchliche Blätter*, 1928.
———. *Vom Sinn und Wert der Familienforschung: Bruchstücke aus einem Vortrag, gehalten im Sommer 1926 auf dem Fürsorgelehrgang in Birthälm*. Hermannstadt: Verein der Kinderfrohen, 1929.
———. "Von Aufartung und Entartung der Völker." *Volk und Heimat* 1 (1939): 13–18.
———. *Von der Bedeutung der Gattenwahl für Wesen und Schicksal der Kinder*. Hermannstadt: Honterus, 1934. Also published in *Volk und Rasse—Volk und Raum* 1 (1934); *Aufbau: Monatschrift* 2 (1934).
———. "Von einst und jetzt und nicht erkannten Führerpflichten." *Blätter für Kinderfrohe* 1–2 (1929): 6–7.
———. "Vorehelich gezeugte Kinder in einigen Gemeinden des Schenker Kirchenbezirks." *Siebenbürgische Vierteljahresschrift* 3–4 (1938): 175–84. Also published in *Volk und Rasse—Volk und Raum* 9 (1938).
———. "Vorehelich gezeugte Kinder in Roseln." *Korrespondenzblatt für siebenbürgische Landeskunde* 49 (1926): 67–68.
———. "Vorschläge zur Neuregelung der Familienzulagen." In *Kleine Aufsätze*, by Alfred Csallner. Hermannstadt: Selbstverlag, 1928. Also published in *Kirchliche Blätter* 34 (1927).
———. "Was soll man dazu sagen?" *OB*, September 30, 1933, 2.
———. "Was wir wollen." *Evangelischer Fürsorger* 6 (1927): 9–12.
———. *Werden wir Tüchtiger oder Untüchtiger*. Hermannstadt: n.p., 1937. Also published in *SDT*, February 16, 1937.
———. *Werkverzeichnis von Alfred Csallner*. Manuscript, n.d. Siebenbürgen-Institut Archiv, Ordner C, Alfred Csallner.
———. "Zur Errichtung des Rassenamtes der 'Selbsthilfe.'" *OB*, October 1, 1932, 4.
———. "Zur Frage der Mischehe." *Volk und Heimat* 1 (1937): 3–10.
———. "Zur gesellschaftsbiologischen Untersuchung unseres Ärztestandes." *Medizinische Zeitschrift* 9 (1931): 29.
———. *Zur Verleihung des siebenbürgisch-sächsischen Kulturpreises 1982: Antwort auf die Laudatio von Dr. Ernst Wagner, Erstem Vorsitzendem des Arbeitskreises für Siebenbürgische Landeskunde*. Manuscript, 1982. Siebenbürgen-Institut Archiv, Ordner C, Alfred Csallner.
———. *Zur wirtschaftlichen und sozialen Lage der Siebenbürger Sachsen 1910–1950: Vier Beiträge*, **edited by** Ernst Wagner. Köln; Wien: Böhlau, 1989.
Csallner, Alfred, Heinrich Brandsch, Hermann Büsch, Rüdiger Fritsch, and Hugo Sitzler. *Denkschrift des "Vereins der Kinderfrohen."* Hermannstadt: Honterus, 1928. Also published in *Evangelischer Fürsorger* 5–6 (1928).
Csallner, Alfred, Fritz Fabritius, and Alfred Bonfert. *Organisationsrichtlinien des Landesamtes für Statistik und Sippenwesen*. Schässburg: Friedrich J. Horeth, 1939.
Csallner, Alfred, and Hauptleitung des VdK. "Zum neuen Anfang." *Blätter für Kinderfrohe* 1 (1930): 1.
Csallner, Daniel. "Der Nösner Gau." *Programm des ev. Obergymnasiums und der damit verbundenen Lehranstallten zu Bistritz*. Hermannstadt: n.p., 1865.

Csallner, Elfriede. "'Es ist fast nichts übrig geblieben . . .': Alfred Csallners wissenschaftliche Arbeiten." *Siebenbürgische Zeitung*, April 30, 1975, 2.
Deletant, Dennis. *Hitler's Forgotten Ally: Ion Antonescu and His Regime, Romania 1940–44*. Houndsmill: Palgrave Macmillan, 2006.
Deutsch-Sächische Frauenbund, ed. *Merkblatt für die sächsische Frau*. N.p., 1934.
Dikotter, Frank. *Imperfect Conceptions: Medical Knowledge, Birth Defects, and Eugenics in China*. London: Hurst, 1998.
Dolle, Heinrich. *Aus Not zu Brot*. Hellerau: Hakenkreuz-Verlag, 1918.
———. "Unser Ziel." *Selbsthilfe* 1 (1922): 1.
———. "Wie soll diese Selbsthilfe aussehen? Teil II." *Selbsthilfe* 3 (1922): 2–3.
Eatwell, Roger. *Fascism: A History*. New York: Allen Lane, 1996.
———. "On Defining the 'Fascist Minimum': The Centrality of Ideology." *Journal of Political Ideologies* 3 (1996): 303–19.
E.C. "'Selbsthilfe.'" *Selbsthilfe* 5 (1925): 1.
Eraso, Yolanda. "Biotypology, Endocrinology, and Sterilization: The Practice of Eugenics in the Treatment of Argentinian Women during the 1930s." *Bulletin of the History of Medicine* 4 (2007): 793–822.
Fabritius, Fritz. "10 Gebote für jeden Selbsthilfemann." *Selbsthilfe*, December 15, 1930, 4.
———. "10 Jahre Selbsthilfearbeit." *Selbsthilfe*, May 21, 1932, 1.
———. "10 Jahre Selbsthilfearbeit." *Selbsthilfe*, May 28, 1932, 1–2.
———. "1931." *Selbsthilfe*, January 1, 1931, 1.
———. "An meine Kampf- und Gesinnungsgenossen!" *OB*, July 28, 1934, 1.
———. "Die Arbeit, die unsere Vereinigung leistet, ist aus den folgenden Berichten zu ersehen." *Selbsthilfe* 7 (1927): 1–3.
———. "Arbeits-Dienstpflicht: angepasst unseren siebenbürgischen Verhältnissen." *Selbsthilfe*, June 27, 1931, 3.
———. "Aufbauarbeit." *Selbsthilfe* 2 (1930): 1.
———. "Baugenossenschaft: Teil I." *Selbsthilfe* 5 (1925): 2.
———. "Baugenossenschaft: Teil II." *Selbsthilfe* 6 (1925): 1–2.
———. "Deutsche Volksgenossen." *OB*, September 30, 1933, 3.
———. "Deutsches Schausspiel?" *Selbsthilfe* 11 (1928): 2.
———. "[Obituary] Dr. Carl Wolff." *Selbsthilfe* 24 (1929): 1.
———. "Frontkämpfer heraus!" *Selbsthilfe* 13 (1928): 1.
———. "Führer." *Selbsthilfe*, October 15, 1930, 1.
———. "Heimaterde." *Selbsthilfe* 7 (1927): 1.
———. "Kameraden!" *Selbsthilfe* 3 (1930): 1.
———. "Kameraden!" *Selbsthilfe*, May 1, 1931, 1–3.
———. "Kameraden!" *Selbsthilfe*, June 15, 1931, 1.
———. "Selbsthilfe." *Selbsthilfe* 12 (1928): 2–3.
———. "Selbsthilfe." *Klingsor* 10 (1930): 412–14.
———. "Siedlung und Siedlerschulen." *Selbsthilfe* 8 (1927): 1–2.
———. "Tue ein jeder nichts als seine Pflicht!" *Selbsthilfe* 10 (1928): 1.
———. "Unser nächstes Ziel." *Selbsthilfe*, June 20, 1931, 1–2.
———. "Was wir wollen." *Selbsthilfe*, December 15, 1930, 1–2.
———. "Welche Arbeiten hat die Selbsthilfe bisher geleistet?" *Selbsthilfe* 6 (1925): 5.
———. "Weltanschauungsfragen." *Selbsthilfe* 8 (1930): 1–3.
———. "Wer hilft mit!?" *Selbsthilfe*, July 18, 1931, 6.
———. "Wer soll den neuen Bischof wählen?" *OB*, August 27, 1932, 4.
Fahlbusch, Michael. *Wissenschaft im Dienst der nationalsozialistischen Politik? Die "Volksdeutschen Forschungsgemeinschaften" von 1931–1945*. Baden-Baden: Nomos, 1999.
———. *Wo der deutsche . . . ist, ist Deutschland!: Die Stiftung für deutsche Volks- und Kulturbodenforschung in Leipzig 1920–1933*. Bochum: Brockmeyer, 1994.

Feldman, Matthew, Marius Turda, and Tudor Georgescu, eds. *Clerical Fascism in Interwar Europe.* Abingdon: Routledge, 2008.
FH [Holzträger, Fritz]. "Die deutsche Volksforschung in Siebenbürgen und ihre Eingliederung in den Neuaufbau der deutschen Volksforschung Rumäniens." *Volksforschung* 4 (1942): 300–302.
Fischer-Galati, Stephen. *Twentieth Century Rumania.* New York, London: Columbia University Press, 1970.
Frank, Richard. "Das Gesundheitswesen unserer Volksgruppe." *Der Arzt im Osten* 1–2 (1943): 7–8.
Gabel, Ingrid. "'Wir wollen frei sein, um zu dienen': Der 'Freie sächsische Frauenbund' von seinen Anfängen bis 1933." *ZfSL* 1 (1999): 67–78.
Gentile, Emilio. "Fascism as Political Religion." *Journal of Contemporary History* 2–3 (1990): 229–51.
———. *The Origins of Fascist Ideology, 1918–1925: The First Complete Study of the Origins of Italian Fascism.* New York: Enigma Books, 2005.
Georgescu, Tudor. "Ethnic Minorities and the Eugenic Promise: The Transylvanian Saxon Experiment with National Renewal in Interwar Romania." *European Review of History* 17, no. 6 (2010): 861–80.
———. "The Eugenic Fortress: Alfred Csallner and the Transylvanian Saxon Eugenic Discourse in Interwar Romania." In *Hygiene, Health and Eugenics in Southeast Europe to 1945*, edited by Marius Turda, Sevasti Trubeta, and Christian Promitzer, 351–84. Budapest: Central European University Press, 2011.
———. "Saxon Eugenics in Transylvania." In *The History of East-Central European Eugenics, 1900–1945: Texts and Commentaries*, edited by Marius Turda, 550-94. London: Bloomsbury, 2015.
Georgescu, Vlad. *The Romanians: A History.* Athens: Ohio State University Press, 1990.
Gündisch, Konrad, and Mathias Beer. *Siebenbürgen und die Siebenbürger Sachsen.* Norneuburg: Langen Müller, 2005.
Glass, Hildrun. "Wer ist ein Deutscher? Anmerkungen zum Selbstverständnis der Rumäniendeutschen (1919–1944)." *HJS* 2 (1997): 15–29.
———. *Zerbrochene Nachbarschaft: das deutsch-judische Verhaltnis in Rumanien, 1918–1938.* München: R. Oldenbourg, 1996.
Glondys, Viktor. "Samaritergeist." *Selbsthilfe*, October 28, 1931, 2–3.
Gregor, A. James. *Phoenix: Fascism in Our Time.* New Brunswick: Transaction Publishers, 1999.
Greisling, H. "Rasse und Volkstum." *Selbsthilfe* 6 (1925): 2–5.
Griffin, Roger. "Fascism's New Faces (and Facelessness) in the 'Post-Fascist' Epoch." In *West and East: An International Debate on Concepts and Cases in the Comparative Study of the Extreme Right*, edited by Roger Griffin, Werner Loh, and Andreas Umland, 29–68. Stuttgart: Ibidem Verlag, 2006.
———. *Modernism and Fascism.* Houndmills: Palgrave Macmillan, 2007.
———. *The Nature of Fascism.* London: Pinter Publishers, 1991.
———. "Tunnel Visions and Mysterious Trees: Modernist Projects of National and Racial Regeneration, 1880–1939." In *"Blood and Homeland": Eugenics and Racial Nationalism in Central and Southeast Europe, 1900–1940*, **edited by** Marius Turda and Paul Weindling, 417–56. Budapest: Central European University Press, 2007.
Griffin, Roger, and Matthew Feldman, eds. *Fascism: Critical Concepts in Political Science.* Abingdon: Routledge, 2003.
Griffin, Roger, Werner Loh, and Andreas Umland, eds. *Fascism Past and Present, West and East: An International Debate on Concepts and Cases in the Comparative Study of the Extreme Right.* Stuttgart: Ibidem Verlag, 2006.
Gust, Waldemar. "Das bessarabische Deutschtum unter neuer Führung." *OB*, May 5, 1934, 1.
———. "Letzter Appell." *OB*, September 30, 1933, 1.
———. "Der wahre Schuldige [Hans Otto Roth]." *OB*, July 28, 1934, 1.
———. "Zur Verschiebung des Sachsentages." *OB*, July 1, 1933, 1.
Gyurgyevich, Julius Ernst. "Aus gesegneter Ernte: Zum schriftstellerischen Schaffen des Dr. Heinrich Siegmund." *Evangelischer Fürsorger* 8 (1937): 31–32.
———. "[Obituary] Dr. Heinrich Siegmund." *SDT*, July 27, 1937, 5.

———. "Vorarbeiten zu einem Schriftenverzeichnis des Landeskonsistorialrats Dr. med. Heinrich Siegmund." *Medizinische Zeitschrift* 10 (1937): 315–26.
Hager, Wilhelm. "Die völkische Nachbarschaft: Eine notwendige Gemeinschaftsform des Siedlungsdeutschen." *OB*, July 1, 1933, 3–4.
———. "[Obituary] Zum Tode Dr. Heinrich Siegmunds." *Klingsor* 8 (1937): 310.
Hager, Wilhelm, and Viktor Weindel. "Deutsche Ärzte!" *Medizinische Zeitschrift* 11 (1940): 283.
Harasser, Albert. Review of "Die Mischehen in den siebenbürgisch-sächsischen Städten und Märkten," by Alfred Csallner. *Archiv für Rassen- und Gesellschafts-Biologie* 6 (1938): 545–46.
———. Review of *Die volksbiologische Forschung unter den Siebenbürger Sachsen*, by Alfred Csallner. *Archiv für Rassen- und Gesellschafts-Biologie* 5 (1940): 393–96.
———. Review of "Völkische Mischehe im Banat," by Matz Hoffmann. *Archiv für Rassen- und Gesellschafts-Biologie* 6 (1938): 543.
Haslinger, Peter. "From a Democracy of Alliance to National Socialist Adherence: The Saxons of Transylvania 1918–1933." Review of *Politische Strukturen und Strömungen bei den Siebenbürger Sachsen 1919–1933*, by Harald Roth. April 2000. http://www.h-net.org/reviews/showrev.php?id=4041.
Haynes, Rebecca. "Germany and the Establishment of the Romanian National Legionary State, September 1940." *Slavonic and East European Review* 4 (1999): 700–72.
———. Introduction to *Occasional Papers in Romanian Studies, No. 2: Papers Presented at the 4th and 5th Romanian Studies Days, SSEES, University of London, March 1996 and February 1997*, edited by Rebecca Haynes, 1–17. London: School of Slavonic and East European Studies, University of London, 1998.
———. "A New 'Greater Romania'? Romanian Claims to the Serbian Banat in 1941." *Central Europe* 2 (2005): 99–120.
———. "Reluctant Allies? Iuliu Maniu and Corneliu Zelea Codreanu against King Carol II of Romania." *Slavonic and East European Review* 85, no. 1 (2007): 105–34.
———. "The Rise of the Political Right in Interwar Romania." Paper presented at the conference "Greater Romania's National Projects: Ideological Dilemmas, Ethnic Classification, and Political Instrumentalization of Ethnic Identities," Oxford Brookes University, April 10–12, 2008. http://www.pulse-project.org/node/77.
———. *Romanian Policy towards Germany, 1936–40*. Houndsmill: Macmillan, 2000.
———. "Work Camps, Commerce, and the Education of the 'New Man' in the Romanian Legionary Movement." *The Historical Journal* 4 (2008): 943–67.
Heim, Susanne, Carola Sachse, and Mark Walker, eds. *The Kaiser Wilhelm Society under National Socialism*. Cambridge: Cambridge University Press, 2009.
Hermann, Albert. "Die Arbeitsschule." *Klingsor* 1 (1932): 15–23.
———. "Die deutsche Jugendbewegung." *Klingsor* 5 (1924): 189–94.
———. *Die deutschen Bauern des Burzenlandes*. Jena: Fischer, 1937.
———. "Die deutschen Bauern des Burzenlandes: Vorbericht über das Ergebnis einer anthropologischen Untersuchung." *Klingsor* 9 (1936): 351–55.
———. "Volksbiologie und Rassenkunde der Burzenländer Sachsen mit Ausblick auf die Siebenbürger Sachsen Überhaupt." *Jahrbuch für Auslandsdeutsche Sippenkunde* 2 (1937): 88–96.
Hermann-Mueller, Grete. *Ze Spet: Volkstaeck aen 3. Afzaejen*. Hermannstadt: Honterus, 1929.
Hienz, Hermann. "Hermann, Dr. phil Albert." In *Schriftsteller-Lexikon der Siebenbuerger Deutschen*. Vol. VII, edited by Hermann Hienz, 117–21. Köln: Boehlau, 2000.
Hillgruber, Andreas. *Hitler, König Carol und Marschall Antonescu; die deutsch-rumänischen Beziehungen, 1938–1944*. Wiesbaden: F. Steiner, 1954.
Hitchins, Keith. *Rumania, 1866–1947*. Oxford: Clarendon Press, 1994.
Hockl, Hans. "Die Nationswerdung der Deutschen in Rumänien, Teil I." *OB*, June 24, 1933, 1.
———. "Die Nationswerdung der Deutschen in Rumänien, Teil II." *OB*, July 1, 1933, 2.
Hoffmann, Matz. "Völkische Mischehen im Banat: Ein Beitrag zur Frage der Umvolkung." *Auslandsdeutsche Volksforschung* 2 (1938): 219–27.
Horváh, Franz. "Die Rumäniendeutschen und der Nationalsozialismus." Review of *Die Gleich-*

schaltung der deutschen Volksgruppe in Rumänien und das "Dritte Reich," 1941–1944, by Johann Böhm. March 2004. http://www.h-net.org/reviews/showpdf.php?id=9001.
Hügel, Eckhard. "Möglichkeiten und Aufgaben der Rassen- und Erbforschung in Siebenbürgen." *Siebenbürgische Vierteljahresschrift* 1 (1936): 7–9.
———. "Rassenforschung und ärztlicher Beruf." *Der Arzt im Osten* 7–8 (1942): 143–47.
———. "Rassenforschung und Volksgruppe." *Deutsche Forschung im Südosten* 1 (1942): 107–14.
———. "Rassenpolitische Gesichtspunkte aus der Geschichte der Deutschen in Rumänien." *Deutsche Forschung im Südosten* 3 (1943): 397–439.
———. Review of *Fürsorge als völkische Selbstbehauptung*, by Paul Collmer. *Siebenbürgische Vierteljahrsschrift* 60 (1937): 239–40.
———. "Volkspflege—Bauernbewegung—Arbeitslager." *Volk im Osten* 22 (1936): 786–95.
Ioanid, Radu. *The Sword of the Archangel: Fascist Ideology in Romania*. New York: Columbia University Press, 1990.
Iordachi, Constantin. *Charisma, Politics, and Violence: The Legion of the "Archangel Michael" in Interwar Romania*. Trondheim: Trondheim Studies on East European Cultures & Societies, 2004.
———, ed. *Comparative Fascist Studies*. London: Routledge, 2009.
———. "Comparative Fascist Studies: An introduction." In *Comparative Fascist Studies*, edited by Constantin Iordachi, 1–50. London: Routledge, 2009.
Jekelius, Ernst. "Wir wollen sein ein ewig Volk von Brüdern." *Selbsthilfe*, August 22, 1931, 1–2.
Jickeli, Karl. "Mädchenerziehung und Rassenhygiene." *Die Karpathen* 2 (1911): 424–28.
Jickeli, Otto Fritz. "Kameraden der NSDR!" OB, October 7, 1933, 3.
———. "NSDR oder 'Eiserne Front'? Rede von Otto Fritz Jickeli anlässlich der Kundgebung der NSDR in Hermannstadt am 7 Juni 1933." *OB*, June 17, 1933, 3–5.
———. *Unser Weg zur Erneuerung des deutschen Volkes in Rumänien*. Hermannstadt: Krafft & Drotleff, 1935.
K. "Kampf!" *OB*, June 17, 1933, 1.
Kallis, Aristotle. *The Fascism Reader*. New York: Routledge, 2003.
Kallis, Aristotle, and António Costa Pinto, eds. *Rethinking Fascism and Dictatorship in Europe*. Basingstoke: Routledge, 2014.
Killyen, Hansgeorg von. "Der Einfluss der Sexualwissenschaft auf das Weltbild der siebenbürgisch-sächsischen Ärzteschaft." *Siebenbürgische Semesterblätter* 6 (1992): 165–68.
———. "Neue Daten zu Studien- und Wirkungsorten siebenbürgisch-sächsischer Ärzte im Zeitraum 1870–1936." *Orvostörténeti Közlemények* 3–4 (2006): 165–79.
———. "Die Rolle der siebenbürgisch-sächsischen Vereine in der Förderung des Gesundheitswesens im Zeitraum 1880–1914." *Orvostörténeti Közlemények* 1–4 (2002): 111–21.
———. "Zur Geschichte des Siebenbürgisch-Sächsischen Ärztevereinwesens (1887–1944)." In *Naturwissenschaftliche Forschungen über Siebenbürgen*. Vol. IV, edited by Heinz Heltmann, 27–65. Köln: Böhlau, 1991.
———. "Zur Geschichte des siebenbürgisch-sächsischen Fürsorgewessens." *ZfSL* 8 (1985): 163–72.
———. "Zur Geschichte der siebenbürgisch-sächsischen Kinderschutzbewegung und der Säuglingspflegerinnen-Schule in Hermannstadt." *ZfSL* 1 (1998): 26–37.
Klausen, Susanne. *Race, Maternity, and Politics of Birth Control in South Africa, 1910–1939*. Basingstoke: Palgrave Macmillan, 2004.
Klein, Karl Kurt. "Geschichtswissenschaft und naturwissenschaftliche Pseudohistorie." *Siebenbürgische Vierteljahresschrift* 4 (1932): 315–28.
Klein, Wilhelm. "... lies Kirchenteppiche pfänden: Aufenthalt im Gefängnis: Pfarrer Alfred Csallner/Als der 'Kinderfrohe' bekannt geworden/Erinnerungen an bedeutende Siebenbürger Sachsen in einer Portraitfolge von Dr. jur. Wilhelm Klein [1888–1976] [Geschrieben Juli 1973, aus dem Nachlass herausgegeben von Wolfgang Wittstock]." *Neuer Weg* 29 (1991): 4.
Klingler, Fritz. "Die Bevölkerungsbewegung des Volkes im Banat vor der Ansiedlung bis heute: Der Geburtenrückgang, dessen Ursachen und die Abhilfe." *Der Aufbau: Monatschrift* 3–4 (1934): 1–14.

Lammert, Emmerich. "Rasse als geschichtsbildender Faktor: Ein Vortrag von Prof. Eugen Fischer in Temeschburg." *Medizinische Zeitschrift* 12 (1941): 290–95.
Larsen, Stein Ugelvik, Bernt Hagtvet, and Jan Petter Myklebust. *Who Were the Fascists: Social Roots of European Fascism*. Bergen: Universitetsforlaget, 1980.
Lebzelter, Viktor. *Eine rassenkundliche Übersichtsaufnahme des Burgenlandes*. Wien: Selbstverlag der Anthropologischen Gesellschaft, 1937.
Lebzelter, Viktor, Josef Wastl, and Anna Sittenberger. *Ein Beitrag zur Rassenkunde der Siebenbürger Sachsen*. Hermannstadt: Krafft & Drotleff, 1936.
Lenz, Fritz. *Die biologischen Grundlagen der Erziehung*. München: Lehmann, 1927.
Liebig, Sabine. "Siebenbürger Sächsinnen zur Zeit des Nationalsozialismus." *ZfSL* 1 (1998): 57–76.
Linger, Heinrich. "Sitte, Moral und Volksreinheit bei Heinrich Siegmund." *Siebenbürgische Semesterblätter* 2 (1992): 169–72.
Livezeanu, Irina. *Cultural Politics in Greater Romania: Regionalism, Nation Building, and Ethnic Struggle, 1918–1930*. Ithaca: Cornell University Press, 1995.
Lumans, Valdis. *Himmler's Auxiliaries: The Volksdeutsche Mittelstelle and the German National Minorities of Europe, 1933–1945*. Chapel Hill: University of North Carolina Press, 1993.
Martini, J. "Pfarrversammlung im Unterwald." *Kirchliche Blätter* 46 (1927): 501–502.
Mathias, Adolf. "Der Aufbau des Amtes für Volksgesundheit." *Der Arzt im Osten* 1–2 (1943): 5–7.
May, Walter. *Ein Jahr Partei: Gestalt und Wandlung der Deutschen Volksgruppe in Rumaenien*. Hermannstadt: Krafft und Drottleff, 1941.
McArthur, Marilyn. *Zum Identitätenwechsel der Siebenbürger Sachsen: Eine Kulturanthropologische Studie*. Köln, Wien: Böhlau, 1990.
Meltzl, Oskar von. "Statistik der Sächsischen Landbevölkerung in Siebenbürgen." *Archiv des Vereins für Siebenbürgische Landeskunde* 2–3 (1885): 213–510.
Miege, Wolfgang. *Das Dritte Reich und die Deutsche Volksgruppe in Rumänien 1933–38: Ein Beitrag zur nationalsozialistischen Volkstumspolitik*. Bern: Lang, 1972.
Milata, Paul. "Anmerkungen zu einem Plagiatsverdacht: Zu Johann Böhms 'Hitlers Vasallen der Deutschen Volksgruppe in Rumänien vor und nach 1945'." *ZfSL* 30 (2007): 163–67.
Möckel, Andreas. "Kleinsächsisch oder Alldeutsch? Zum Selbstverständnis der Siebenbürger Sachsen von 1867 bis 1933." In *Siebenbürgen zwischen den beiden Weltkriegen*, edited by Walter König, 129–48. Köln: Böhlau, 1994.
———. *Umkämpfte Volkskirche: Leben und Wirken des evangelischen Pfarrers Konrad Möckel (1892–1965)*. Köln: Böhlau, 2011.
Möller, Karl von. "Arbeitsordnung für die erste Kulturamtswalter-Schulung des Kulturamtes der Deutschen in Rumänien." *Der Aufbau: Monatschrift* 2 (1934): 18–19.
———. "Der Aufbau." *Der Aufbau: Monatschrift* 8–9 (1935): 1.
———. "Behelfe für rassische Aufklärungsarbeit." *Der Aufbau: Monatschrift* 1 (1934): 21.
———. "Erbbiologischer und eugenischer Wegweiser für jedermann." *Aufbau: Monatschrift* 6 (1934): 7.
———. "Erbsünden." *Aufbau: Monatschrift* 6 (1934): 6–7.
———. *Zwischen Hitler, Stalin und Antonescu: Rumäniendeutsche in der Waffen-SS*. Köln: Böhlau, 2007.
Morrissey, Christof N. "National Socialism and the Ethnic Germans of Romania." Review of *Die Gleichschaltung der deutschen Volksgruppe in Rumänien und das "Dritte Reich," 1941–1944*, by Johann Böhm. November 2004. http://www.h-net.org/reviews/showrev.php?id=9923.
Müller, Georg. "Die Aufgaben unserer Kirche in der Gesundheitsfuehrung unseres Volkes: Teil I." *Evangelischer Fürsorger* 2 (1939): 5–7.
———. "Die Aufgaben unserer Kirche in der Gesundheitsfuehrung unseres Volkes: Teil II." *Evangelischer Fürsorger* 3 (1939): 9–10.
———. "Das Landeskirchliche Fürsorgeausstellung 'Volksgesundheit.'" *Evangelischer Fürsorger* 6 (1939): 21.
———. "Zum Abschied." *Evangelischer Fürsorger* 11–12 (1940): 41.
Neugeboren, Emil. "Rundschreiben der Gauleitung Siebenbürgen and alle Kreis und Ortsgruppenleite." *OB*, April 21, 1934, 7.

Nolte, Ernst. *The Three Faces of Fascism: Action Française, Italian Fascism, National Socialism*. London: Weidenfeld and Nicolson, 1965.
Oberth, Hermann. "Ist Siegmunds 'Deutschen Dämmerung' unwissenschaftlich?" *Siebenbürgische Vierteljahresschrift* 4 (1932): 302–15.
Offner, Robert. "Zur Geschichte des siebenbürgisch-ungarischen Ärztevereinwesens." *ZfSL* 1 (1995): 23–34.
Orend, Misch. "Der Nationalsozialismus und die 'Intellektuellen.'" *OB*, March 24, 1934, 1.
Payne, Stanley G. *A History of Fascism, 1914–1945*. Basingstoke: Routledge, 1995.
Pendas, Devin O. *The Frankfurt Auschwitz Trial, 1963–1965: Genocide, History, and the Limits of the Law*. New York: Cambridge University Press, 2006.
Popa, Klaus. *Akten um die "Deutsche Volksgruppe in Rumaenien" 1937–1945: Eine Auswahl*. Frankfurt am Main: Peter Lang, 2006.
———. "Deutsche Forschung im Südosten." In *Handbuch der völkischen Wissenschaften: Personen—Institutionen—Forschungsprogramme—Stiftungen*, edited by Ingo Haar and Michael Fahlbusch. München: Sauer, 2008.
———. "Erfüllung des NS-Ungeistes: Bibliographisch-biographische Portraits aus der 'Deutschen Volksgruppe in Rumänien' 1940–1944: Teil IV [Alfred Pomarius]." *HJS* 2 (2006): 86–89.
———. "Erfüllung des NS-Ungeistes: Bibliographisch-biographische Portraits aus der 'Deutschen Volksgruppe in Rumänien' 1940–1944: Teil V [Misch Orend]." *HJS* 1 (2007): 49–59.
———. "Neokonservative Propadandaschrift: Zu: Paul Milita: Zwischen Hitler, Stalin und Antonescu: Rumäniendeutsche in der Waffen-SS." *HJS* 2 (2007): 14.
———. Review of *Die Gleichschaltung der deutschen Volksgruppe in Rumänien und das "Dritte Reich," 1941–1944*, by Johann Böhm. August 2003. http://www.h-net.org/reviews/showrev.php?id=18189.
———. *Die Rumäniendeutschen zwischen Demokratie und Diktatur: Der politische Nachlass von Hans Otto Roth 1919–1951*. Frankfurt am Main: Peter Lang, 2003.
Proctor, Robert. *Racial Hygiene: Medicine under the Nazis*. Cambridge: Harvard University Press, 1988.
Quandt, Viktor. *Nachbarschaft, die überschaubare Einheit mit Herz: Ein kurzer Überblick über Sinn und Wandel des Nachbarschaftswesens der Siebenbürger Sachsen in Vergangenheit und Gegenwart*. Wien: Kramer Verlag, n.d. [ca. 1971].
———. "Der Wiedererwecker der nachbarlichen Hilfe." Newspaper cutout.
R. "Die Nachbarschaften und die Selbsthilfe." *Selbsthilfe* 11 (1928): 1–2.
R.E.F. "Der Kampf der Geister." *Selbsthilfe* 10 (1930): 1–2.
Rehner, J. *Dem Andenken Dr. Heinrich Siegmunds weiland Grosstemplers von Rumäniens Grosslaube 2 (deutsch) des I.O.G.T.* N.d. [1937].
Reinerth, Karl. "Zu den Innenpolitischen Auseinandersetzungen unter den Deutschen in Rumänien zwischen den beiden Weltkriegen." In *Siebenbürgen zwischen den beiden Weltkriegen*, edited by Walter König, 149–69. Köln: Böhlau, 1994.
———. *Zur politischen Entwicklung der Deutschen in Rumänien 1918–1928: Aus einer siebenbürgisch-sächsischen Sicht*. Köln: Böhlau Verlag, 1994.
Reinerth, Karl, and Fritz Cloos. *Zur Geschichte der Deutschen in Rumänien 1935–1945: Beiträge und Berichte*. Bad Tölz: Arbeitsgemeinschaft für Südostdeutsche Volks- und Heimatforschung, 1988.
Ritter, Ernst. *Das Deutsche Auslandsinstitut in Stuttgart, 1917–1945*. Wiesbaden: Franz Steiner, 1976.
Roberts, David. *The Totalitarian Experiment in Twentieth-Century Europe: Understanding the Poverty of Great Politics*. Basingstoke: Routledge, 2005.
Roemer-Neubner, Meta. "Frauenbewegung und Rassenhygiene." *Die Karpathen* 2 (1911): 538–44.
Rosenauer, Rudolf. "In Memoriam Dr. Heinrich Siegmund." *Evangelischer Fürsorger* 8 (1937): 29–30.
Roth, Harald. *Kleine Geschichte Siebenbürgens*. Köln: Böhlau, 2003.
———, ed. *Minderheit und Nationalstaat: Siebenbürgen seit dem ersten Weltkrieg*. Köln: Böhlau, 1995.
———. "Patterns of Interwar Transylvanian Saxon Identity." Paper presented at the conference "Greater Romania's National Projects: Ideological Dilemmas, Ethnic Classification, and Political Instrumentalization of Ethnic Identities," Oxford Brookes University, April 10–12, 2008. http://www.pulse-project.org/node/72.

———. *Politische Strukturen und Strömungen bei den Siebenbürger Sachsen, 1919–1933*. Köln: Böhlau, 1994.
———. "Wissenschaft zwischen Nationalsozialismus und Stalinismus." In *Südostforschung im Schatten des Dritten Reiches: Institutionen, Inhalte, Personen*, edited by Mathias Beer and Gerhard Seewann, 123–33. Munich: Oldenbourg, 2004.
———. "Zum Wandel der politischen Strukturen bei den Siebenbürger Sachsen 1918 bis 1933." In *Minderheit und Nationalstaat: Siebenbürgen seit dem Ersten Weltkrieg*, edited by Harald Roth, 99–114. Köln: Böhlau, 1995.
Roth, Harald, and Mariana Hausleitner, eds. *Der Einfluss von Faschismus und Nationalsozialismus auf Minderheiten in Ostmittel- und Südosteuropa*. München: IKGS, 2006.
Rr. "Zur Krise in unserem Volk: Teil I." *Selbsthilfe*, July 11, 1931, 2–3.
———. "Zur Krise in unserem Volk: Teil II." *Selbsthilfe*, July 18, 1931, 1–2.
S. "Gautagung in Hermannstadt: Teil I." *Selbsthilfe*, June 27, 1931, 3–4.
———. "Gautagung in Hermannstadt: Teil II." *Selbsthilfe*, July 18, 1931, 3–4.
———. "Sie bitten ums Wort, Herr Pfarrer? Unsere Antwort an Herrn Pfarrer Hermann." *Selbsthilfe*, June 1, 1931, 1–2.
Sächsischer Ärzteverein, ed. *Gutachten des sächsischen Ärztevereins über die zukünftiger Gestaltung des sächsischen Gesundheitswesens, verhandelt und angenommen in seiner Hauptversammlung vom 6. Juli 1919 in Mediasch*. 1919.
Scheeser, Dr. "7 Juni und 'Selbsthilfe!'" *Selbsthilfe*, September 26, 1931, 2.
Scheiner, Herwart. "Selbst Entlarvt!" *OB*, July 28, 1934, 2.
Schlesak, Dieter. *Capesius, der Auschwitzapoetheker*. Bonn: Dietz, 2006.
Schmidt, Andreas. "Gründung der Nationalsozialistischen Arbeiterpartei der Deutschen Volksgruppe in Rumänien." In *Nationalsozialistischer Volkstumskampf*, by Andreas Schmidt, 19–21. Hermannstadt: Krafft und Drotleff, 1941.
———. "Organisationsrichtlinien des Amtes für Organisation und Gemeinschaftsbildung der Deutschen Volksgruppe in Rumänien." *Arbeits-Nachrichten*, December 21, 1940, 3.
———. "Der vorläufige Arbeitsrahmen der Volksorganisation der deutschen Volksgruppe in Rumänien." *Arbeits-Nachrichten*, December 21, 1940, 4.
Schmuhl, Hans-Walter, ed. *The Kaiser-Wilhelm-Institute for Anthropology, Human Heredity and Eugenics, 1927–1945: Crossing Boundaries*. Dordrecht: Springer, 2007.
———. *Rassenforschung an Kaiser-Wilhelm-Instituten vor und nach 1933*. Goettingen: Wallstein Verlag, 2003.
Schubert, Hans-Achim. *Nachbarschaft: Entfremdung und Protest*. Freiburg und München, 1977.
———. *Nachbarschaft und Modernisierung*. Köln: Böhlau, 1980.
———. "Wilhelm Schunn's 'Völkische' Nachbarschaften—Wunsch und Wirklichkeit." In *Siebenbürgen zwischen den beiden Weltkriegen*, edited by Walter König, 169–82. Köln: Böhlau, 1994.
Schuller, G.A. "Die älteste Matrikel von Reussdörfchen." *Kirchliche Blätter*, 1913.
———. "Aus der ältesten Galter Matrikel." *Kirchliche Blätter*, 1928.
Schullerus, Adolf. *Siebenbürgisch sächsisches Wörterbuch: Vol. I, A C*. Berlin, Leipzig: Walter de Gruyter & Co., 1908.
Schunn, Wilhelm. "Die Gautagung in Schässburg." *Selbsthilfe*, November 15, 1930, 3–4.
———. "Die Nachbarschaften als geeignete Lebensform für aussendeutsche Volksgruppen." *Volk und Heimat* 6–7 (1938): 122–23.
———. *Die Nachbarschaften der Deutschen in Rumänien*. Hermannstadt: Krafft & Drotleff, 1937.
———. *Das neue Merkblatt über Ziele und Einrichtungen der Selbsthilfe*. Hermannstadt: Honterus, 1931.
———. *Schulungsbüchlein für Amtswalter des Nachbarschaftswesens nachdem Stande der Arbeit im Herbst 1939*. Hermannstadt: Krafft & Drotleff, 1939.
———. "Der Start zu einem Hohen Ziel: Bericht des Beauftragten für den volksorganischen Aufbau." In *Ein Jahr Volksgemeinschaft der Deutschen in Rumänien unter Fritz Fabritius*, by Helmut Wolff, 47–67. Hermannstadt: Krafft & Drotleff, 1936.
———. *Weg und Feinde der NEDR*. Hermannstadt: Schlosser, 1934.

Schunn, Wilhelm, and Oskar Pastior. *Die Ehrung des Kinderreichtums bei den Deutschen in Rumänien.* Krafft & Drotleff, 1940.

Schuster, Oskar, ed. *Epoche der Entscheidungen: Die Siebenbürger Sachsen im 20 Jahrhundert.* Köln; Wien: Böhlau, 1984.

Shaper, Erna. "Volk schaff dir Raum." *Selbsthilfe* 8 (1927): 3–4.

Siebenbürgisch-deutscher Ärzteverein, ed. "Die rechtlichen Grundlagen des siebenbürgisch-deutschen Ärztevereins." *Medizinische Zeitschrift* 3 (1933): 7–19.

Siegmund, Heinrich. "Ansprache zur Eröffnung der landeskirch. Fürsorgeausstellung 'Volksgesundheit.'" *Evangelischer Fürsorger* 5 (1937): 17–19.

———. "Bericht des Stadtphysikus über seinen Besuch der ‚Internationalen Hygiene-Ausstellung' in Dresden." *Mediascher Wochenblatt,* March 15, 1912.

———. "Bericht über die im Zusammenhange mit der Fürsorgewoche abgehaltenen Veranstaltungen." *Evangelischer Fürsorger* 7 (1930): 74–77.

———. *Deutschen-Dämmerung in Siebenbürgen.* Hermannstadt: Honterus, 1931.

———. "Der drohende Volkstod der Siebenbürger Sachsen infolge der Kinderbeschränkung." *Deutsche Erde,* 1913, 42–45.

———. "Fernwirkungen." *Evangelischer Fürsorger* 1 (1937): 2–4.

———. "Füherer durch die landeskirchliche Fürsorgeausstellung 'Volksgesundheit,' Teil I." *Evangelischer Fürsorger* 2–3 (1937): 5–7.

———. "Füherer durch die landeskirchliche Fürsorgeausstellung 'Volksgesundheit,' Teil II." *Evangelischer Fürsorger* 4 (1937): 13–15.

———. "Füherer durch die landeskirchliche Fürsorgeausstellung 'Volksgesundheit,' Teil III." *Evangelischer Fürsorger* 5 (1937): 19–20.

———. *Merkblatt über die Ziele der evangelischen Fürsorge: Fürsorge-Blätter der evangelischen Landeskirche A.B. in Siebenbürgen.* Hermannstadt: Honterus, 1922.

———. "Rassengesundheitspflege." *Evangelischer Fürsorger* 2 (1931): 5–7.

———. "Der Raumsinn des sächsischen Volkes." *Die Karpathen* 5 (1910): 142–47.

———. "Sächsische Volksmehrung und Frauenbewegung, Teil I." *Die Karpathen* 23 (1913): 731–34.

———. "Sächsische Volksmehrung und Frauenbewegung, Teil II." *Die Karpathen* 24 (1913): 750–58.

———, ed. *Sächsisches Wehr und Mehrbuch: Ein Volksbuch.* Mediasch: G.U. Reissenberger, 1922.

———. "Völkische Sittengebote: Teil I." *Evangelischer Fürsorger* 1 (1928): 2–3.

———. "Völkische Sittengebote: Teil II." *Evangelischer Fürsorger* 2 (1928): 12–15.

———. *Was kann und soll die säschsische Siedlungstätigkeit leisten?* GA Reissenberger, 1913, 1.

———. "Die wichtigste Aufgabe der landeskirchlichen Fuersorge." *Evangelischer Fürsorger* 2 (1933): 5–6.

———. *Ziel und Einrichtung des Bodenschutz-Vereins.* Hermannstadt: Heinrich Metzer, n.d.

———. *Zur sächsischen Rassenhygiene.* Hermannstadt: Drotleff, 1901. Also published in *Medizinische Zeitschrift* 9 (1937): 273–81.

Siegmund, Heinrich, Michael Englisch, and Rudolf Schuster, eds. *Sächsisches Wehr und Mehrbuch: Ein Volksbuch.* Mediasch: Selbstverlag, 1914.

Siemens, Hermann Werner. *Grundzüge der Vererbungslehre, der Rassenhygiene und der Bevölkerungspolitik.* München: Lehmann, 1926.

Solonari, Vladimir. "Visions of Ethnic Purity in Greater Romania." Paper presented at the conference "Greater Romania's National Projects: Ideological Dilemmas, Ethnic Classification, and Political Instrumentalization of Ethnic Identities," Oxford Brookes University, April 10–12, 2008. http://www.pulse-project.org/node/83.

Stepan, Nancy Leys. *The Hour of Eugenics: Race, Gender, and Nation in Latin America.* Ithaca: Cornell University Press, 1991.

Swanson, John. "Nation, Volk, Minderheit, Volksgruppe: Die deutsche Minderheit in Ungarn in den Begriffskämpfen der Zwischenkriegsära." *Zeitschrift für Ostmitteleuropa-Forschung,* 55 (2006): 526–47.

———. "The Second World War and Its Aftermath: Ethnic German Communities in the East." In *The Germans and the East*, edited by Charles Ingrao and Franz Szabo, 347–61. West Lafayette: Purdue University Press, 2008.

Szelényi, Balázs. "From Minority to *Übermensch*: The Social Roots of Ethnic Conflict in the German Diaspora of Hungary, Romania, and Slovakia." *Past and Present* 1 (2007): 215–51.

T. "Unser Programm." *Selbsthilfe*, August 1, 1931, 1–3.

Teschler-Nicola, Maria. "*Volksdeutsche* and Racial Anthropology in Interwar Vienna: The 'Marienfeld Project.'" In *"Blood and Homeland": Eugenics and Racial Nationalism in Central and Southeast Europe, 1900–1940*, edited by Marius Turda and Paul Weindling, 55–82. Budapest: Central European University Press, 2007.

Teutsch, Grete. "Dr. Heinrich Siegmund und die Frauenbewegung." *Die Karpathen* 2 (1913): 40–42.

Thomsen, Andreas. *Der Völker Vergehen und Werden: Grundlagen einer allgemeinen Völkerpolitik*. Leipzig: Voigtländer, 1925.

Trașcă, Ottmar. "Die Deutsche Volksgruppe in Rumänien und die Ereignisse vom 23. August 1944 im Spiegel eines unveröffentlichten Manuskripts." *ZfSL* 2 (2011): 186–228.

Turda, Marius. "From Craniology to Serology: Racial Anthropology in Interwar Hungary and Romania." *Journal of the History of Behavioral Sciences* 43, no. 4 (2007): 361–77.

———, ed. *The History of Eugenics in East-Central Europe, 1880-1945: Texts and Commentaries*. London: Bloomsbury, 2015.

———. *The Idea of National Superiority in Central Europe, 1880–1918*. Lampeter: The Edwin Mellen Press, 2004.

———. "In Pursuit of Greater Hungary: Eugenic Ideas of Social and Biological Improvement, 1940–1941." *The Journal of Modern History* 85, no. 3 (2013): 558–91.

———. "In Search of Racial Types: Soldiers and the Anthropological Mapping of the Romanian Nation, 1914–1944." *Patterns of Prejudice* 47, no. 1 (2013): 1–21.

———. *Modernism and Eugenics*. Houndsmill: Palgrave Macmillan, 2010.

———. "Nation and Race in Interwar Romania." Paper presented at the conference "Greater Romania's National Projects: Ideological Dilemmas, Ethnic Classification, and Political Instrumentalization of Ethnic Identities," Oxford Brookes University, April 10–12, 2008, http://www.pulse-project.org/node/75.

———. "The Nation as Object: Race, Blood, and Biopolitics in Interwar Romania." *Slavic Review* 66, no. 3 (2007): 413–41.

———. "'To End the Degeneration of a Nation': Debates on Eugenic Sterilization in Interwar Romania." *Medical History* 53, no. 1 (2009): 77–104.

———. *Eugenics and Nation in Early 20th Century Hungary*. Houndsmill: Palgrave Macmillan, 2014.

Turda, Marius, and Aaron Gillette. *Latin Eugenics in Comparative Perspective*. Bloomsbury Academic, 2014.

Turda, Marius, Christian Promitzer, and Sevasti Trubeta, eds. *Health, Hygiene and Eugenics in Southeastern Europe to 1945*. Budapest: Central European University Press, 2011.

Turda, Marius, and Paul Weindling, eds. *"Blood and Homeland": Eugenics and Racial Nationalism in Central and Southeast Europe, 1900–1940*. Budapest: Central European University Press, 2007.

V.C.J. "Eine Erwiderung auf die Rede von Dr. H.O. Roth am 3 April in Bistritz." *OB*, April 21, 1934, 1.

Vlaicu, Monika, ed. *120 de ani de arhivă publică în Transilvania/120 Jahre öffentliches Archiv in Siebenbürgen*. Hermannstadt: Editura Tenis-Club-Sen, 1996.

Wagner, E. "Ansprache des Gauführers [Burzenland] Dr. Wagner, Heldsdorf, in der Mitgliederversammlung vom 21 Juli 1930 in Kronstadt." *Selbsthilfe* 9 (1930): 2.

Wa[gner, Ernst]. "Wirken zum Wohle der Gemeinschaft: Alfred Csallner 95 Jahre alt." *Siebenbuergische Zeitung*, May 15, 1990, 11.

Waber, Gustav. "Gedanken zur Reinhaltung der Rasse." *Selbsthilfe* 1 (1930): 2–3.

———. "Ortsgruppe Kronstadt: Teil I." *Selbsthilfe* 20 (1929): 4–6.

———. "Ortsgruppe Kronstadt: Teil II." *Selbsthilfe* 21 (1929): 6–8.

Wagner, Ernst. *Geschichte der Siebenbürger Sachsen: Ein Überblick*. Innsbruck: Wort und Welt Verlag, 1981.

———. "Heinrich Sigmund und die 'Volksbiologische' Forschung in der Zwischenkriegszeit." *ZfSL* 2 (1983): 177–86.

———. *Quellen zur Geschichte der Siebenbürger Sachsen: 1191–1975*. Köln, Wien: Böhlau, 1976.

Weber, Andreas. "Ein offenes Wort." *OB*, May 20, 1933, 2.

Weber, Eugen. *Varieties of Fascism: Doctrines of Revolution in the Twentieth Century*. New York: Van Nostrand Reinhold Company, 1964.

Weber, Georg. "Zum Struktur- und Funktionswandel der siebenbürgisch-sächsischen Nachbarschaft in Rumänien und Westdeutschland: Eingrenzung, Begriffe, Institutionen." *Ethnologia Europaea* 12 (1981): 131–61.

Weindel, Viktor. "Die Eingliederung des in das Forschungsinstitut der Deutschen Volksgruppe in Rumänien." *Mitteilungen der Arbeitsgemeinschaft für Naturwissenschaften* 91–92 (1941–42): 5–9.

———. "Zum Geleit!" *Mitteilungen der Arbeitsgemeinschaft für Naturwissenschaften* 91–92 (1941–42): 1.

Weindling, Paul. *Health, Race, and German Politics between National Unification and Nazism 1870–1945*. Cambridge: Cambridge University Press, 1989.

———. *Nazi Medicine and the Nuremberg Trials: From Medical War Crimes to Informed Consent*. Basingstoke: Palgrave Macmillan, 2004.

———. "Regenerating Mitteleuropa: Methodological and Theoretical Reflections on Eugenics, Hygiene, and Health Politics in Central and Southeast Europe." In *Health, Hygiene and Eugenics in Southeastern Europe to 1945*, edited by Christian Promitzer, Marius Turda, and Sevasti Trubeta. Budapest–New York: CEU Press, 2011.

Weingart, Peter, Jürgen Kroll, and Kurt Bayertz. *Rasse, Blut und Gene: Geschichte der Eugenik und Rassenhygiene in Deutschland*. Frankfurt: Suhrkamp, 1992.

Weiss, Sheila Faith. *Race, Hygiene, and National Efficiency: The Eugenics of Wilhelm Schallmayer*. Berkeley; Los Angeles: University of California Press, 1987.

Wichern, Heinrich. *Sexualität und Bevölkerungspolitik*. Schwerin: Friedrich Bahn, 1927.

Wien, Andreas Ulrich. ."'Entjudung' und Nationalsozialismus als Ziel des Religionsunterrichts: Zum Lehrplan für den Religionsunterricht an deutschen Schulen 1942." *ZfSL* 30 (2007): 65–77.

———. "'Ich rufe alle auf, sich in die Heimat einzukrallen und entschlossen zu sein, hier zu bleiben': Dr. Hans Otto Roth als Politiker und Landeskirchenkurator der Evangelischen Landeskirche AB in Rumänien." In *Beiträge zur ostdeutschen Kirchengeschichte*. Vol. VII, edited by Peter Maser and Christian-Erdmann Schott, 176–88. Münster: Verein für ostdeutsche Kirchengeschichte, 2005.

———. *Kirchenleitung über dem Abgrund. Bischof Friedrich Müller vor den Herausforderungen durch Minderheitenexistenz, Nationalsozialismus und Kommunismus*. Köln: Böhlau, 1998.

Wokalek, Franz. "Der Arzt als Treuhänder der Volksgesundheit." *Medizinische Zeitschrift* 8 (1941): 168–74.

———. "Nationalsozialistische Volksgesundheitsführung." *Der Arzt im Osten* 2 (1942): 33–34.

———. *Statistische Untersuchungen über die Häufigkeit erblicher Augenleiden*. Berlin: Pfau, 1939.

———. "Die Volksgesundheitliche Lage und Betreuung der Volksgruppe." In *Jahrbuch der Deutschen Volksgruppe in Rumänien, 1942*, 175–80. Hermannstadt: Krafft und Drotleff, 1941.

———. "Zum Geleit!" *Der Arzt im Osten* 1 (1942): 1.

Wolf, Siegbert. "Zur Konstituierung des Verbandes der Deutschen in Rumänien." *Selbsthilfe*, September 26, 1931, 4.

Wolff, Helmut. *Ein Jahr Volksgemeinschaft der Deutschen in Rumänien unter Fritz Fabritius*. Hermannstadt: Krafft & Drotleff, 1936.

WR. "Was ist Arbeitsdienst? Der Arbeitsdienst die Erziehungsschule zur wahren Volksgemeinschaft." *OB*, May 20, 1933, 4.

Zach, Cornelius. Review of *Die Deutschen in Rumänien und das Dritte Reich 1933–1940*, by Johann Böhm. *ZfSL* 2 (1999): 299–301.

INDEX OF NAMES

INDEX OF PLACES